AN ANALYTICAL CALCULUS

VOLUME IV

AN ANALYTICAL CALCULUS

FOR SCHOOL AND UNIVERSITY

BY

E. A. MAXWELL

Fellow of Queens' College, Cambridge

VOLUME IV

DIFFERENTIAL EQUATIONS AND ANALYSIS

CAMBRIDGE

AT THE UNIVERSITY PRESS

1968

CAMBRIDGE UNIVERSITY PRESS
Cambridge, New York, Melbourne, Madrid, Cape Town, Singapore, São Paulo, Delhi

Cambridge University Press
The Edinburgh Building, Cambridge CB2 8RU, UK

Published in the United States of America by Cambridge University Press, New York

www.cambridge.org
Information on this title: www.cambridge.org/9780521056991

First published 1957
Reprinted 1968
This digitally printed version 2008

A catalogue record for this publication is available from the British Library

ISBN 978-0-521-05699-1 hardback
ISBN 978-0-521-09041-4 paperback

CONTENTS

SECTION 1

ORDINARY DIFFERENTIAL EQUATIONS

CHAPTER XIX: EQUATIONS OF THE FORM $y'=f(x, y)$

CHAPTER XX: LINEAR DIFFERENTIAL EQUATIONS; GENERAL PROPERTIES

CHAPTER XXI: THE LINEAR DIFFERENTIAL EQUATION WITH CONSTANT COEFFICIENTS

CHAPTER XXII: THE LINEAR DIFFERENTIAL EQUATION WITH CONSTANT COEFFICIENTS; ALTERNATIVE METHOD

SECTION 2

THE DEFINITION OF FUNCTIONS BY INFINITE SERIES AND INTEGRALS

CHAPTER XXIII: THE CONVERGENCE OF SERIES

CHAPTER XXIV: THE DEFINITION OF FUNCTIONS BY SERIES

Chapter XXV: POWER SERIES

Chapter XXVI: THE DEFINITION OF FUNCTIONS BY INTEGRALS

Chapter XXVII: THE SOLUTION OF DIFFERENTIAL EQUATIONS IN TERMS OF INFINITE SERIES

CHAPTER XXVIII: FOURIER SERIES

SECTION 3

LAPLACE'S EQUATION AND RELATED EQUATIONS

CHAPTER XXIX: THE TRANSFORMATION OF LAPLACE'S EQUATION

CHAPTER XXX: 'LAPLACE' EQUATIONS

CHAPTER XXX (*cont.*)

CHAPTER XXXI: SPHERICAL HARMONICS

PREFACE

I would express my gratitude to my pupils T. H. Gossling, G. H. Mitchell, D. W. G. Moore and A. Robinson for help with the examples and for suggestions for improvements to the text—an example of the real inversion of normal functions.

As the fourth volume brings this work to its completion, I record my very deep indebtedness to Dr Sheila M. Edmonds, of Newnham College, Cambridge, who read the whole manuscript most searchingly. Those who know her personally will realize how much trouble she has taken; others would believe it impossible. I am most grateful to her.

I am very happy to say how much I appreciate the cheerful and inspiring help given by the staff of the Cambridge University Press over a number of years.

E. A. M.

QUEENS' COLLEGE, CAMBRIDGE
May 1957

NOTE ON THE SECOND IMPRESSION

I am grateful for kindly comments and for suggestions for improvements. I am particularly indebted to Dr L. E. Clarke and to Professor R. L. Goodstein. Changes are not substantial, but attention should be called to pages 115, 133 and 257—9.

E. A. M.

June 1962

INTRODUCTION

This volume, as originally planned, was intended to conclude the whole work with a review, chiefly in differential equations, of such standard theory of the calculus as could be exhibited without a detailed study of analysis. I soon found, however, that analytical requirements kept penetrating and could not be kept out without loss of intellectual honesty. The volume is therefore much longer than I intended, and includes, substantially, a whole freshman's course of analysis, and more in addition. Nevertheless, my aim remained to keep the exposition as simple as possible within clearly stated limitations.

The theme of the volume is the differential equation and its solution; and it is hoped that the treatment shows how the processes of solution demand extended definitions of functions (for example, series and integrals) together with a technique (analysis) for studying and controlling their behaviour. The aim is not so much to elaborate the detailed properties of such fresh functions, as to instil *methods* which the student can apply or, better, adapt himself when faced later with the need for extending his mathematical vocabulary.

The work is, in essence, familiar, but it ought perhaps to be remarked that there are a number of points where the details vary from standard practice.

SECTION 1

ORDINARY DIFFERENTIAL EQUATIONS

There is a large field of mathematics, especially in its application to physical problems, in which the aim, ultimate or intermediate, is to express one variable y as a function of another variable x. In simple cases the restatement of a given problem in mathematical language may give the relationship quickly; in others, the solution may come only after a strenuous struggle with algebraic or trigonometrical equations.

What we have to consider now is the possibility that the language may, in the first instance, involve not only the variables themselves, but also differential coefficients, ordinary or partial. In this section we confine our attention to the ordinary differential coefficients dy/dx, d^2y/dx^2, and so on.

For example, a curve may be known to have the property that, referred to a given system of rectangular Cartesian coordinates, the rate of change of the gradient at any point $P(x, y)$ is equal to the square of the distance of P from the y-axis. The problem of identifying the curve begins in mathematical language with the equation

$$\frac{d}{dx}\left(\frac{dy}{dx}\right) = x^2,$$

or

$$\frac{d^2y}{dx^2} = x^2.$$

In this simple case we can go further at once and 'solve' the equation. By integration,

$$\frac{dy}{dx} = \tfrac{1}{3}x^3 + A,$$

where A may have any arbitrary constant value; and a further integration gives the 'solution'

$$y = \tfrac{1}{12}x^4 + Ax + B,$$

where B is a second arbitrary constant. Thus there is a *family* of curves with the given property, and individual members of the family are picked out by the particular values of A, B; for instance, if we know that the curve passes through the two points (0, 0) and (1, 0), then substitution of these values in the solution gives the relations

$$B=0, \quad A=-\tfrac{1}{12},$$

so that the curve is

$$y=\tfrac{1}{12}x^4-\tfrac{1}{12}x.$$

It is worthy of note that this general solution for the relation with the *second* differential coefficient contains *two* arbitrary constants.

A relation like $y''=x^2$, involving differential coefficients, is called a DIFFERENTIAL EQUATION; an equation is ORDINARY or PARTIAL according as the differential coefficients are ordinary (as throughout this section) or partial. The ORDER of the equation is the order of the highest differential coefficient contained in it. The process of expressing y as a function of x is called SOLVING the differential equation, and, for order n, the solution may be expected to contain n arbitrary constants whose evaluation, when required, depends on 'conditions' beyond the equation itself. Whether any particular equation can be solved at all is a big problem on which we do not enter; our attention is directed towards examples which experience has proved to be soluble.

We are dealing with a subject which is one of the largest in mathematics, and for a fully detailed account the reader must pass to the text-books specially devoted to it. The aim of this section is to explain the principles which underlie the processes of solution, with sufficient detail to enable the general student to tackle the problems which he is most likely to encounter at the present stage of his work.

CHAPTER XIX

EQUATIONS OF THE FORM $y'=f(x, y)$

The solution of a differential equation necessarily begins with a recognition of its *form*. The present chapter deals with equations in which y' is expressed as a single-valued function $f(x,y)$ of x and y. The treatment varies according to the form of the function, and the work which follows is designed to exhibit a number of typical cases.

1. Geometrical interpretation. If the two variables x, y are regarded as the rectangular Cartesian coordinates of a point P, the equation

$$\frac{dy}{dx}=f(x,y)$$

serves to specify a definite gradient at P. A solution appears as a relation (with an arbitrary constant) connecting x, y, and may be interpreted as the equation of a curve, which is the locus of a point P moving so that its direction at P has given gradient $f(x,y)$. The presence of the arbitrary constant is reflected in the arbitrariness in the choice of a starting-point which, once selected, determines the particular curve of the system.

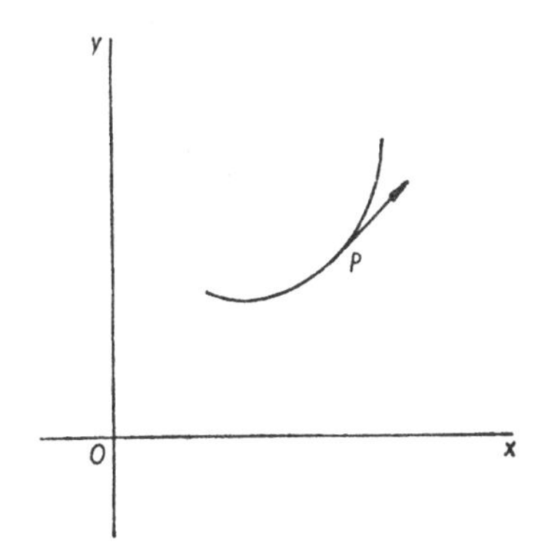

Fig. 147

2. The equation when $f(x, y)$ is a function (i) of x only, (ii) of y only.

(i) If $f(x, y) \equiv F(x)$, the equation is

$$\frac{dy}{dx} = F(x),$$

and the solution is $$y = \int F(x)\, dx + C,$$

where C is an arbitrary constant.

(ii) If $f(x, y) \equiv G(y)$, the equation is

$$\frac{dy}{dx} = G(y),$$

and the solution is $$x = \int \frac{dy}{G(y)} + C,$$

where C is an arbitrary constant.

3. Variables separable. If

$$f(x, y) \equiv u(x)\, v(y),$$

the product of a function of x only with a function of y only, the equation is

$$\frac{dy}{dx} = u(x)\, v(y),$$

and the solution is $$\int \frac{dy}{v(y)} = \int u(x)\, dx + C,$$

where C is an arbitrary constant. The equation has been *separated* into a part involving x only and a part involving y only.

EXAMPLES I

Solve the following differential equations:

1. $\dfrac{dy}{dx} = 1 + x^2$.
2. $\dfrac{dy}{dx} = 1 + y^2$.
3. $\dfrac{dy}{dx} = \cot x \tan y$.
4. $\dfrac{dy}{dx} = 1 + x + y + xy$.

Solve the following differential equations under the conditions stated:

5. $\dfrac{dy}{dx}=(1+x)^2$, given that $y=0$ when $x=0$.

6. $\dfrac{dy}{dx}=\cos y \cot y$, given that $y=0$ when $x=3$.

7. $e^x\dfrac{dy}{dx}=xy^2$, given that $y=1$ when $x=1$.

8. $(1+x)\dfrac{dy}{dx}+1+y=0$, given that $y=8$ when $x=2$.

4. Homogeneous equation.

If

$$f(x,y)\equiv g\left(\frac{y}{x}\right),$$

a function of the quotient y/x, the equation is

$$\frac{dy}{dx}=g\left(\frac{y}{x}\right).$$

Such an equation is called HOMOGENEOUS. It can be solved by reducing it first to the 'variables separable' type:

Make the substitution

$$y=zx,$$

so that

$$\frac{dy}{dx}=z+x\frac{dz}{dx}.$$

The equation is

$$z+x\frac{dz}{dx}=g(z),$$

or*

$$\frac{dx}{x}=\frac{dz}{g(z)-z},$$

so that

$$\log x=\int\frac{dz}{g(z)-z}+C,$$

where C is an arbitrary constant.

ILLUSTRATION 1. *To solve the equation*

$$\frac{dy}{dx}=\frac{xy}{x^2+y^2}.$$

* It is customary, and convenient, to use differentials freely in work of this kind.

Write
$$y=zx,$$
so that
$$\frac{dy}{dx}=z+x\frac{dz}{dx}.$$
The equation is
$$z+x\frac{dz}{dx}=\frac{z}{1+z^2},$$
or
$$x\frac{dz}{dx}=\frac{z}{1+z^2}-z$$
$$=\frac{-z^3}{1+z^2}.$$
Hence
$$\frac{dx}{x}+\frac{(1+z^2)\,dz}{z^3}=0,$$
or
$$\frac{dx}{x}+\frac{dz}{z^3}+\frac{dz}{z}=0,$$
so that
$$\log x-\frac{1}{2z^2}+\log z=\text{constant},$$
or
$$\log zx=\frac{1}{2z^2}+\text{constant},$$
or
$$\log y=\frac{x^2}{2y^2}+\text{constant}.$$
The solution may therefore be expressed in the form
$$y=Ae^{(x^2/2y^2)},$$
where A is an arbitrary constant.

ILLUSTRATION 2. *To solve the equation*
$$\frac{dy}{dx}+\frac{3x-18y-15}{22x-4y+18}=0.$$

We include an example which is not, in the first instance, of homogeneous type in order to show the kind of treatment which some equations require to bring them to standard form.

Make the substitution
$$x=u+a, \quad y=v+b,$$

where u, v are the new variables and a, b constants. Then

$$\frac{dv}{du}=\frac{dy}{dx}=-\frac{3x-18y-15}{22x-4y+18}$$

$$=-\frac{3u-18v+(3a-18b-15)}{22u-4v+(22a-4b+18)}.$$

Choose a, b so that $\quad 3a-18b-15=0,$

$$22a-4b+18=0,$$

or

$$a=-1, \quad b=-1.$$

Hence

$$\frac{dv}{du}=-\frac{3u-18v}{22u-4v}$$

$$=-\frac{3-18(v/u)}{22-4(v/u)}.$$

The equation is now of homogeneous form, and we make the substitution

$$v=zu,$$

so that

$$z+u\frac{dz}{du}=-\frac{3-18z}{22-4z},$$

or

$$u\frac{dz}{du}=\frac{4z^2-4z-3}{22-4z}=\frac{(2z-3)(2z+1)}{22-4z},$$

or

$$\frac{du}{u}=\frac{(22-4z)\,dz}{(2z-3)(2z+1)}$$

$$=\frac{4dz}{2z-3}-\frac{6dz}{2z+1}.$$

Integrating, we have

$$\log u=2\log(2z-3)-3\log(2z+1)+\text{constant},$$

or

$$u=\frac{C(2z-3)^2}{(2z+1)^3},$$

where C is an arbitrary constant. Hence

$$u(2z+1)^3=C(2z-3)^2,$$

or

$$(2v+u)^3=C(2v-3u)^2,$$

or, since $u=x+1$, $v=y+1$,

$$(2y+x+3)^3=C(2y-3x-1)^2.$$

EXAMPLES II

Solve the differential equations:

1. $\dfrac{dy}{dx} = \dfrac{x+y}{x-y}$.

2. $\dfrac{dy}{dx} = \dfrac{x+y+2}{x-y}$.

3. $\dfrac{dy}{dx} = \dfrac{y}{x+y}$.

4. $\dfrac{dy}{dx} = \left(\dfrac{x+y}{x}\right)^2$.

5. $2y\dfrac{dy}{dx} = \dfrac{x+y^2}{x+4y^2}$, by the substitution $y^2 = vx$.

6. $\dfrac{dy}{dx} + \dfrac{y^3+x}{y^2(y^3-x)} = 0$, by appropriate substitution.

7. $\sec^2 y\dfrac{dy}{dx} = \dfrac{x\cos y + 2\sin y}{2x\cos y + \sin y}$, by appropriate substitution.

8. $\dfrac{dy}{dx} + \dfrac{e^{x-y}-2}{2+e^{y-x}} = 0$.

5. Orthogonal trajectories. The solution, with arbitrary constant, of the equation

$$\frac{dy}{dx} = f(x, y)$$

defines, for differing values of the constant, the curves of a family F_1 with the property that the gradient of the member through a point $P(x, y)$ has the value $f(x, y)$. It may happen that there exists a second family F_2 of curves, having the property that, at every point P in the plane, the tangent to the curve of F_2 through P is perpendicular to the tangent to the curve of F_1 through P. The curves of the family F_2 are then called the ORTHOGONAL TRAJECTORIES of the curves of F_1.

Since the gradient of the curve of F_2 through P is $-1/f(x, y)$, the differential equation for the curves of F_2 is

$$\frac{dy}{dx} = \frac{-1}{f(x, y)}.$$

NOTE. *Orthogonal trajectories in polar coordinates.* A differential equation

$$\frac{dr}{d\theta} = F(r, \theta)$$

defines a system of curves referred to polar coordinates r, θ. It is easy to deduce, from the results given in Volume II, pp. 109–10, that the differential equation for the orthogonal trajectories is

$$-\frac{1}{r}\frac{dr}{d\theta}=\frac{r}{F(r,\theta)}.$$

ILLUSTRATION 3. *To find the orthogonal trajectories of the rectangular hyperbolas*

$$x^2-y^2=C,$$

where C is a parameter varying from hyperbola to hyperbola.

The curves of the system satisfy the differential equation

$$x-y\frac{dy}{dx}=0,$$

or

$$\frac{dy}{dx}=\frac{x}{y}.$$

Hence their orthogonal trajectories satisfy the equation

$$\frac{dy}{dx}=-\frac{y}{x},$$

or

$$x\frac{dy}{dx}+y=0.$$

Integrating, we have the equation

$$xy=A,$$

which, for varying A, defines a second family of rectangular hyperbolas.

EXAMPLES III

Find the orthogonal trajectories of the families of curves:

1. $x^2-4y^2=A$.
2. $x^2-y^2+2x=A$.
3. $\sin x \cosh y=A$.
4. $x\cos y-y\sin y=Ae^{-x}$.
5. $x^4-6x^2y^2+y^4=A$.
6. $x^3+x^2-3xy^2-y^2=A$.

REVISION EXAMPLES XV

Solve the differential equations:

1. $2(1-xy)\dfrac{dy}{dx} = y^2.$

2. $\dfrac{1}{2x}\dfrac{dy}{dx} + \dfrac{x+y}{x^2+y^2} = 0.$

3. $\dfrac{dy}{dx} = -\dfrac{ax+hy+g}{hx+by+f}.$

4. $\dfrac{dy}{dx} = \dfrac{1}{2y}\dfrac{4x+3y^2-1}{3x+4y^2+1}.$

5. $\dfrac{dy}{dx} = \dfrac{1-x^2-y}{1+x+y^2}.$

6. $\dfrac{dy}{dx} = \dfrac{3x-5y-9}{2x-4y-8}.$

7. $(x+y-1)\dfrac{dy}{dx} = x+y+1.$

8. $\dfrac{dy}{dx} + \dfrac{x^2+3y^2}{3x^2+y^2} = 0.$

9. $(y^3-xy)\dfrac{dy}{dx} = 1.$

10. $\left(\dfrac{dy}{dx}\right)^2 = (x-y)\dfrac{dy}{dx} + xy.$

11. $\left(\dfrac{dy}{dx}\right)^2 + (\sin x + \cos x)\, y\dfrac{dy}{dx} + \tfrac{1}{2}y^2 \sin 2x = 0.$

12. $\dfrac{dy}{dx} + y^2 = 1$, where $y=0$ when $x=0$.

13. $v\dfrac{dv}{dx} = g - kv^2$, given that $v=0$ when $x=0$.

14. $x^2\dfrac{dy}{dx} = y^2 + 2xy.$

15. $x\dfrac{dy}{dx} - y^2 = xy^2.$

16. $(x^2-2xy)\dfrac{dy}{dx} = y^2 - 2xy.$

17. $(5x-y+1)\dfrac{dy}{dx} + x - 5y + 5 = 0.$

18. $\dfrac{dy}{dx} = \dfrac{x \tanh y}{x^2+1}$, given that $y=1$ when $x=0$.

19. $x^2\dfrac{dy}{dx} = y(x+y)$, given that $y=-1$ when $x=1$.

20. $\cos x\dfrac{dy}{dx} - y\sin x = 1.$

21. $\dfrac{dy}{dx} = \dfrac{x+y+1}{x-2y+3}.$

22. $\dfrac{dy}{dx} + y\cot x = \sin^3 x.$

23. $2\dfrac{dy}{dx} = \dfrac{y}{x} + \dfrac{y^2}{x^2}.$

24. $e^{-x}\cos x\dfrac{dy}{dx}-ye^{-x}\sin x=x$, given that $y=0$ when $x=0$.

25. $y^2+(xy+x^2)\dfrac{dy}{dx}=0$, given that $y=3$ when $x=1$.

26. Obtain the differential equation satisfied by the family of curves

$$y=Cx+x\log x,$$

where C is a variable parameter.

Find the differential equation of the orthogonal family and integrate it.

27. Obtain a differential equation satisfied by the family of curves

$$y^3-3x^2y=a^3,$$

where a is a variable parameter.

Find the differential equation of the orthogonal family, and integrate it.

28. A family of curves is given by

$$e^{y-x}=\lambda(y+x),$$

where λ is a variable parameter. Find the differential equation satisfied by this family and show that there is an orthogonal family, and that it consists of parabolas.

29. Find the differential equation of the first order satisfied by the family of curves

$$y^2=\frac{x^3-a^3}{3x},$$

where a is a variable parameter.

Find the equation to the family of curves that is orthogonal to the above family.

30. Find the orthogonal trajectories of the families of curves

(i) $y(x^2-y^2)=C(x^2+y^2)$,

(ii) $r=C(1+\cos\theta)$.

CHAPTER XX

LINEAR DIFFERENTIAL EQUATIONS; GENERAL PROPERTIES

[The beginner is advised to read this chapter fairly quickly to gain familiarity with the outline and the terminology. Too much stress should not be put at first on the more abstract ideas; but they will appear later as important to a real understanding of the manipulations.]

1. The linear operator. We turn now to the LINEAR DIFFERENTIAL EQUATION

$$P_0(x)\frac{d^n y}{dx^n} + P_1(x)\frac{d^{n-1}y}{dx^{n-1}} + \ldots + P_{n-1}(x)\frac{dy}{dx} + P_n(x)\,y = Q(x),$$

in which y and its differential coefficients occur linearly. The corresponding equation when $Q(x)$ is replaced by zero is called the COMPLEMENTARY EQUATION.

The left-hand side may be expressed more concisely by using a symbol D to denote the operation of differentiation with respect to the current variable x. Thus

$$Dy \equiv \frac{dy}{dx}, \quad Dx^3 = 3x^2, \quad D\sin x = \cos x,$$

and so on. It is important to observe that the operator D acts on a term that comes *after* it; for example,

$$x^2 Dx^5 = x^2(5x^4) = 5x^6,$$

but

$$x^5 Dx^2 = x^5(2x) = 2x^6.$$

By an obvious process of induction, we write

$$D^2 y \equiv D(Dy) = D\left(\frac{dy}{dx}\right) = \frac{d^2 y}{dx^2},$$

$$D^3 y \equiv D(D^2 y) = D\left(\frac{d^2 y}{dx^2}\right) = \frac{d^3 y}{dx^3},$$

and so on; then the given equation appears in the form

$$\{P_0(x)\,D^n + P_1(x)\,D^{n-1} + \ldots + P_{n-1}(x)\,D + P_n(x)\}\,y = Q(x).$$

The expression in brackets { } is called a LINEAR OPERATOR acting on y.

The PRODUCT of two linear operators

$$\{P_0(x)\,D^n + \ldots + P_n(x)\}$$

and
$$\{Q_0(x)\,D^m + \ldots + P_m(x)\}$$

is defined by the chain

$$\{P_0(x)\,D^n + \ldots + P_n(x)\}\{Q_0(x)\,D^m + \ldots + Q_m(x)\}\,y \equiv \{P_0(x)\,D^n + \ldots + P_n(x)\}\,v,$$

where
$$v \equiv \{Q_0(x)\,D^m + \ldots + Q_m(x)\}\,y.$$

The product of more than two operators is defined similarly by induction.

The order of the factors in a product is very important; the two expressions

$$\{P_0(x)\,D^n + \ldots + P_n(x)\}\{Q_0(x)\,D^m + \ldots + Q_m(x)\}\,y$$

and
$$\{Q_0(x)\,D^m + \ldots + Q_m(x)\}\{P_0(x)\,D^n + \ldots + P_n(x)\}\,y$$

are usually quite distinct. For example

$$\begin{aligned}(xD+1)(x^2D+1)\,y &= (xD+1)(x^2y'+y)\\ &= x\frac{d}{dx}(x^2y'+y) + (x^2y'+y)\\ &= x(x^2y''+2xy'+y') + (x^2y'+y)\\ &= x^3y'' + (3x^2+x)\,y' + y,\end{aligned}$$

whereas

$$\begin{aligned}(x^2D+1)(xD+1)\,y &= (x^2D+1)(xy'+y)\\ &= x^2\frac{d}{dx}(xy'+y) + (xy'+y)\\ &= x^2(xy''+2y') + (xy'+y)\\ &= x^3y'' + (2x^2+x)\,y' + y.\end{aligned}$$

An operator like $\quad P_0(x)\,D^n + \ldots + P_n(x)$

is often abbreviated to the symbol

$$L(D).$$

If $L(D)$ and $M(D)$ are the two operators just mentioned, the two products are $L(D)\,M(D)$ and $M(D)\,L(D)$; when acting on y they will yield the two expressions

$$L(D)\,M(D)\,y, \quad M(D)\,L(D)\,y.$$

EXAMPLES I

Evaluate the following expressions:

1. $(D^2-3D+2)\,e^{4x}$.
2. $(D^2-5D+6)\,e^{3x}$.
3. $(D^4+4D^2+4)\,x\sin 2x$.
4. $(\sin x\,.\,D+\cos x)\cos^2 x$.
5. $(\sin x\,.\,D\cos x)-(\cos x\,.\,D\sin x)$.
6. $(x^2D^2x^3)-(x^3D^2x^2)$.
7. $(D^2-2D+1)\,ye^{2x}$.
8. $(D^3-3D^2+3D-1)\,ye^x$.
9. $(xD+2)(xD+3)\,y$.
10. $(D-1)(xD-2)\,e^x$.
11. $(\cos x\,.\,D+\sin x)(\sin x\,.\,D+\cos x)\cos x$.
12. $(e^x D+e^{-x})(e^{-x} D+e^x)\,e^{3x}$.

2. Linearly dependent functions. The n functions

$$u_1(x), \quad u_2(x), \quad \ldots, \quad u_n(x)$$

are called LINEARLY DEPENDENT if there exists a linear identity

$$A_1u_1(x)+A_2u_2(x)+\ldots+A_nu_n(x)\equiv 0$$

with constants $A_1, A_2, \ldots, A_n$ not all zero.

For example, the functions

$$x, \quad x^2, \quad x+x^2$$

are linearly dependent, whereas the functions

$$x, \quad x^2, \quad x^3$$

are not.

To prove that, *if the functions $u_1, u_2, \ldots, u_n$ are linearly dependent, then the determinant $W(x)$ defined by the relation*

$$W(x)\equiv\begin{vmatrix} u_1 & u_2 & \ldots & u_n \\ u_1' & u_2' & \ldots & u_n' \\ \ldots & \ldots & \ldots & \ldots \\ u_1^{(n-1)} & u_2^{(n-1)} & \ldots & u_n^{(n-1)} \end{vmatrix}$$

is identically zero.

(For the three functions x, x^2, $x+x^2$, the identical relation is

$$1.x+1.x^2+(-1).(x+x^2)\equiv 0,$$

and the determinant $W(x)$ is

$$\begin{vmatrix} x & x^2 & x+x^2 \\ 1 & 2x & 1+2x \\ 0 & 2 & 2 \end{vmatrix},$$

which is zero for all values of x.)

If the functions are linearly dependent, there must exist constants $A_1, A_2, \ldots, A_n$ *not all zero*, such that

$$A_1 u_1 \quad + A_2 u_2 \quad + \ldots + A_n u_n \quad \equiv 0.$$

On the assumption that the differential coefficients exist, successive differentiation gives the relations

$$\begin{array}{l} A_1 u_1' \quad + A_2 u_2' \quad + \ldots + A_n u_n' \quad \equiv 0, \\ \ldots\ldots\ldots\ldots\ldots\ldots\ldots\ldots\ldots\ldots \\ A_1 u_1^{(n-1)} + A_2 u_2^{(n-1)} + \ldots + A_n u_n^{(n-1)} \equiv 0. \end{array}$$

Since the constants $A_1, A_2, \ldots, A_n$ are not all zero, the determinant obtained by eliminating them from these n relations must be zero; that is,

$$W(x) \equiv 0.$$

The determinant $W(x)$ is called the WRONSKIAN of the given functions.

ILLUSTRATION 1. *The n functions*

$$e^{a_1 x}, \quad e^{a_2 x}, \quad \ldots, \quad e^{a_n x}$$

are linearly independent if $a_1, a_2, \ldots, a_n$ are all different.

The Wronskian is

$$\begin{vmatrix} e^{a_1 x} & e^{a_2 x} & \ldots & e^{a_n x} \\ a_1 e^{a_1 x} & a_2 e^{a_2 x} & \ldots & a_n e^{a_n x} \\ \ldots & \ldots & \ldots & \ldots \\ a_1^{n-1} e^{a_1 x} & a_2^{n-1} e^{a_2 x} & \ldots & a_n^{n-1} e^{a_n x} \end{vmatrix}$$

$$\equiv e^{(a_1+a_2+\ldots+a_n)x} \begin{vmatrix} 1 & 1 & \ldots & 1 \\ a_1 & a_2 & \ldots & a_n \\ \ldots & \ldots & \ldots & \ldots \\ a_1^{n-1} & a_2^{n-1} & \ldots & a_n^{n-1} \end{vmatrix}$$

$$\equiv \pm e^{(a_1+a_2+\ldots+a_n)x} (a_1-a_2)(a_1-a_3)\ldots(a_{n-1}-a_n).$$

Since this is not zero if $a_1, a_2, \ldots, a_n$ are all different, the n functions are linearly independent.

(Note that we are concerned with LINEAR independence. For example, the functions $e^x, e^{2x}, e^{3x}, e^{4x}$ satisfy the quadratic identity

$$e^x e^{4x} \equiv e^{2x} e^{3x}.)$$

ILLUSTRATION 2. *The n functions*

$$e^{ax}, \quad xe^{ax}, \quad x^2e^{ax}, \quad \ldots, \quad x^{n-1}e^{ax}$$

are linearly independent.

Since an identity

$$A_1 e^{ax} + A_2 x e^{ax} + \ldots + A_n x^{n-1} e^{ax} \equiv 0$$

implies an identity

$$A_1 + A_2 x + \ldots + A_n x^{n-1} \equiv 0,$$

it will suffice to prove that $1, x, \ldots, x^{n-1}$ are linearly independent.

The Wronskian is

$$\begin{vmatrix} 1 & x & x^2 & x^3 & \ldots & x^{n-1} \\ 0 & 1 & 2x & 3x^2 & \ldots & (n-1)\,x^{n-2} \\ 0 & 0 & 2.1 & 3.2x & \ldots & (n-1)(n-2)\,x^{n-3} \\ \ldots & \ldots & \ldots & \ldots & \ldots & \ldots \\ 0 & 0 & 0 & 0 & \ldots & (n-1)! \end{vmatrix} \equiv 1.2!.3!\ldots..(n-1)!.$$

Since this is not zero, the n functions are linearly independent.

3. The complementary function. Consider the linear differential equation

$$L(D)\,y \equiv \{P_0(x)\,D^n + \ldots + P_n(x)\}\,y = Q(x),$$

with complementary equation

$$L(D)\,y = 0.$$

Suppose that we have been able, in any manner, to obtain solutions $y = u_1(x), y = u_2(x), \ldots$ of the complementary equation, so that

$$L(D)\,u_1(x) = 0, \quad L(D)\,u_2(x) = 0, \quad \ldots.$$

Then, by direct substitution in the complementary equation, we find that *the function*

$$y = A_1 u_1(x) + A_2 u_2(x) + \ldots$$

satisfies the equation $$L(D)\,y=0$$

for all sets of values of the constants $A_1, A_2, \ldots$.

In particular, if
$$u_1(x), \quad u_2(x), \quad \ldots, \quad u_n(x)$$

are *precisely n linearly independent solutions* of the complementary equation, then the solution

$$y=A_1u_1(x)+A_2u_2(x)+\ldots+A_nu_n(x)$$

is called the COMPLEMENTARY FUNCTION of the given equation. It contains the n arbitrary constants that might be expected. (Compare p. 2.)

For example, it is easy to verify that each of the functions e^x, e^{2x} satisfies the equation

$$\frac{d^2y}{dx^2}-3\frac{dy}{dx}+2y=0,$$

and so the complementary function of the equation

$$\frac{d^2y}{dx^2}-3\frac{dy}{dx}+2y=4x\sin 3x$$

is $$Ae^x+Be^{2x}.$$

We assume without proof the theorem that EVERY *solution of the complementary equation can be expressed in this way, as a linear combination of any n linearly independent solutions.*

4. Solution by complementary function and particular integral. Suppose that we have been able, in any manner, to find one solution
$$y=U(x)$$

of the *given* equation $$L(D)\,y=Q(x).$$

Such a solution is called a PARTICULAR INTEGRAL.

To prove that, *if* $y=u_1(x)$, $y=u_2(x)$, … *are solutions of the complementary equation and if* $y=U(x)$ *is a particular integral (of the given equation), then the function*

$$y=A_1u_1(x)+A_2u_2(x)+\ldots+U(x)$$

is a solution of the given equation.

With this value of y,

$$L(D)\,y \equiv \{L(D)\,A_1u_1\} + \{L(D)\,A_2u_2\} + \ldots + \{L(D)\,U\}$$
$$\equiv A_1\{L(D)\,u_1\} + A_2\{L(D)\,u_2\} + \ldots + \{L(D)\,U\},$$

since $A_1, A_2, \ldots$ are constants. Thus

$$L(D)\,y \equiv A_1.0 + A_2.0 + \ldots + Q,$$

by definition of $u_1, u_2, \ldots, U$. Hence

$$L(D)\,y = Q.$$

In particular, if $u_1(x), u_2(x), \ldots, u_n(x)$ are precisely n linearly independent functions satisfying the complementary equation, then a solution, with n arbitrary constants, of the given equation is

$$y = A_1u_1(x) + A_2u_2(x) + \ldots + A_nu_n(x) + U(x).$$

We assume without proof that *every* solution of the given equation can be expressed in this way, as the sum of the complementary function and any one particular integral.

(For different choices of particular integral we should require different sets of values of the arbitrary constants $A_1, A_2, \ldots, A_n$.)

ILLUSTRATION 3. *To solve the equation*

$$\frac{d^2y}{dx^2} - 3\frac{dy}{dx} + 2y = 4.$$

The complementary function (compare p. 17) is

$$Ae^x + Be^{2x}.$$

Also it is obvious that a particular solution of the given equation is

$$y = 2.$$

Hence the general solution is

$$y = Ae^x + Be^{2x} + 2.$$

5. The equation of first order. The linear differential equation of the first order is

$$P_0(x)\frac{dy}{dx} + P_1(x)\,y = Q(x),$$

but it is more convenient to divide first by $P_0(x)$ to obtain the form

$$\frac{dy}{dx}+P(x)\,y=Q(x)$$

(for a fresh value of $Q(x)$ on the right-hand side).

This equation can always be solved, subject only to the evaluation in practice of the integrals which appear. We give two methods, of which the first has been foreshadowed in Volume III (p. 54).

FIRST METHOD. *Integrating factor.*

Evaluate first the integral

$$\int P(x)\,dx$$

and then the exponential $e^{\int P(x)\,dx}$, written sometimes $e^{\int P\,dx}$.

The latter expression is an integrating factor of the given equation which, on multiplying by that factor, becomes

$$e^{\int P\,dx}\frac{dy}{dx}+P(x)\,e^{\int P\,dx}\,y=Q(x)\,e^{\int P\,dx},$$

or
$$\frac{d}{dx}\{ye^{\int P\,dx}\}=Q(x)\,e^{\int P\,dx}.$$

Thus
$$ye^{\int P\,dx}=C+\int Q(x)\,e^{\int P\,dx}\,dx,$$

where C is an arbitrary constant. The equation is therefore solved, in the form
$$y=Ce^{-\int P(x)\,dx}+e^{-\int P(x)\,dx}\int Q(x)\,e^{\int P(x)\,dx}\,dx.$$

SECOND METHOD. *Variation of parameters.*

This method can be applied widely and is of general importance.

We begin with the complementary equation

$$\frac{dy}{dx}+P(x)\,y=0,$$

or
$$\frac{dy}{y}+P(x)\,dx=0,$$

and obtain at once its solution

$$y=ae^{-\int P(x)\,dx},$$

where a is an arbitrary constant.

The method consists in taking this solution of the complementary equation as a kind of approximation to the solution of the original equation, but replacing the constant a by a variable z; that is, we make in the given equation the substitution

$$y = ze^{-\int P(x)\,dx}.$$

Thus

$$ye^{\int P(x)\,dx} = z,$$

so that

$$e^{\int P(x)\,dx}\frac{dy}{dx} + P(x)\,e^{\int P(x)\,dx}\,y = \frac{dz}{dx},$$

and the given equation therefore appears in the form

$$\frac{dz}{dx} = Q(x)\,e^{\int P(x)\,dx}.$$

Hence

$$z = \int Q(x)\,e^{\int P(x)\,dx}\,dx + C,$$

where C is an arbitrary constant, so that

$$\begin{aligned} y &= ze^{-\int P(x)\,dx} \\ &= Ce^{-\int P(x)\,dx} + e^{-\int P(x)\,dx}\int Q(x)\,e^{\int P(x)\,dx}\,dx. \end{aligned}$$

In the language of the preceding paragraphs, the complementary function is

$$Ce^{-\int P(x)\,dx},$$

and a particular integral is

$$e^{-\int P(x)\,dx}\int Q(x)\,e^{\int P(x)\,dx}\,dx.$$

ILLUSTRATION 4. *To solve the equation*

$$(1+x)\frac{dy}{dx} + xy = (1+x)^2\,e^{2x}.$$

We use the method of variation of parameters, which is more in line with some of the work that follows.

The complementary equation is

$$(1+x)\frac{dy}{dx} + xy = 0,$$

or

$$\frac{dy}{y} + \frac{x\,dx}{1+x} = 0,$$

or

$$\frac{dy}{y} + \left(1 - \frac{1}{1+x}\right)dx = 0,$$

and its solution is

$$\log y + x - \log(1+x) = \text{constant},$$

or $$y = a(1+x)\,e^{-x},$$

where a is an arbitrary constant. We therefore subject the given equation to the substitution

$$y = z(1+x)\,e^{-x},$$

so that $$\frac{dy}{dx} = \frac{dz}{dx}(1+x)\,e^{-x} - zxe^{-x},$$

and the equation becomes

$$(1+x)^2 e^{-x}\frac{dz}{dx} = (1+x)^2\,e^{2x},$$

or $$\frac{dz}{dx} = e^{3x}.$$

Hence $$z = \tfrac{1}{3}e^{3x} + C,$$

and the solution of the given equation is

$$y = C(1+x)\,e^{-x} + \tfrac{1}{3}(1+x)\,e^{2x}.$$

See Revision Examples XVI, nos. 1–20, pp. 26–7.

6. The equation of second order. The linear differential equation of the second order is

$$P_0(x)\frac{d^2y}{dx^2} + P_1(x)\frac{dy}{dx} + P_2(x)\,y = Q(x).$$

It is not possible to derive a general solution in the straightforward way that we have just done for the equation of the first order. There are, however, a number of occasions when one solution of the *complementary* equation can be 'spotted', and it is then possible to solve the given equation completely. We prove that, if ONE *solution of the complementary equation is known, then the given equation may be solved completely*, subject only to the evaluation of the integrals involved.

Suppose that $$y = u(x)$$

is a known solution of the complementary equation, so that (in abbreviated notation)

$$P_0u'' + P_1u' + P_2u = 0.$$

Remembering that $y = au$ is also a solution for constant a, we use the method of Variation of Parameters, and make the substitution

$$y = zu$$

in the given equation; then

$$y' = z'u + zu',$$

$$y'' = z''u + 2z'u' + zu''.$$

The given equation becomes

$$P_0(z''u + 2z'u' + zu'') + P_1(z'u + zu') + P_2 zu = Q,$$

or $$P_0 uz'' + (2P_0 u' + P_1 u) z' + (P_0 u'' + P_1 u' + P_2 u) z = Q,$$

or $$uz'' + 2u'z' + (P_1/P_0) uz' = Q/P_0$$

on dividing by P_0 and remembering that the coefficient of z is zero.

Multiply* by u. Then

$$\frac{d}{dx}(u^2 z') + (P_1/P_0)(u^2 z') = (Qu/P_0).$$

This equation is linear in the variable $u^2 z'$, and so (p. 19)

$$u^2 z' = Ae^{-\int (P_1/P_0)\,dx} + e^{-\int (P_1/P_0)\,dx} \int \frac{Qu}{P_0} e^{\int (P_1/P_0)\,dx}\,dx,$$

where A is an arbitrary constant.

Divide by u^2 and integrate; then multiply by u. We obtain the solution of the given equation in the form

$$y = Bu + Au \int \frac{e^{-\int (P_1/P_0)\,dx}}{u^2}\,dx + u \int \frac{e^{-\int (P_1/P_0)\,dx}}{u^2} \left\{ \int \frac{Qu}{P_0} e^{\int (P_1/P_0)\,dx}\,dx \right\} dx,$$

where B is a second arbitrary constant.

As with much of this work, it is the *method* and not the formula that should be remembered.

ILLUSTRATION 5. *To solve the equation*

$$(x^2 - 2x) y'' - (x^2 - 2) y' + 2(x - 1) y = (x^2 - 4x + 2) e^x.$$

The complementary equation is

$$(x^2 - 2x) y'' - (x^2 - 2) y' + 2(x - 1) y = 0.$$

* The expression $uz'' + 2u'z'$ becomes, after multiplication by u, the differential coefficient $\frac{d}{dx}(u^2 z')$. The student must learn to 'focus his eyes' to observe such possibilities.

We observe that

$$(x^2-2x)-(x^2-2)+2(x-1)\equiv 0,$$

and so

$$y = ae^x$$

is a solution of the complementary equation (since then we have $y=y'=y''$). We therefore make the substitution

$$y=ze^x,$$

so that

$$y'=z'e^x+ze^x,$$

$$y''=z''e^x+2z'e^x+ze^x,$$

and the given equation becomes

$$(x^2-2x)\,e^x\,(z''+2z'+z)-(x^2-2)\,e^x\,(z'+z)+2(x-1)\,e^x z$$
$$=(x^2-4x+2)\,e^x,$$

or

$$(x^2-2x)\,z''+\{2(x^2-2x)-(x^2-2)\}\,z'=x^2-4x+2,$$

or

$$(x^2-2x)\,z''+(x^2-4x+2)\,z'=x^2-4x+2.$$

Thus

$$(x^2-2x)\,z''+(x^2-4x+2)\,(z'-1)=0,$$

so that

$$\frac{d(z')}{z'-1}+\frac{x^2-4x+2}{x^2-2x}\,dx=0,$$

or

$$\frac{d(z')}{z'-1}+\left\{1-\frac{2x-2}{x^2-2x}\right\}dx=0.$$

Integrating,

$$\log(z'-1)+x-\log(x^2-2x)=\text{constant},$$

or

$$z'-1=A(x^2-2x)\,e^{-x}.$$

Hence

$$z-x=A\int(x^2-2x)\,e^{-x}\,dx+B$$
$$=-Ax^2e^{-x}+B,$$

where A, B are arbitrary constants.

Thus

$$z=B-Ax^2e^{-x}+x,$$

and so the solution of the given equation is

$$y=Be^x-Ax^2+xe^x.$$

7. Solution by factorization of the operator. Suppose the second-order linear differential equation to be divided throughout by the coefficient of y'' and then expressed in the form

$$L(D)\,y \equiv \{D^2 + P_1(x)\,D + P_2(x)\}\,y = Q(x).$$

A solution is obtained if the left-hand side can be factorized in the form

$$\{D + u(x)\}\,\{D + v(x)\}\,y;$$

for, writing

$$\{D + v(x)\}\,y \equiv z,$$

the first-order linear equation

$$\frac{dz}{dx} + u(x)\,z = Q(x)$$

gives z, and then the first-order linear equation

$$\frac{dy}{dx} + v(x)\,y = z$$

gives y. The problem is to find $u(x)$ and $v(x)$.

Since

$$\begin{aligned}\{D + u(x)\}\,\{D + v(x)\}\,y &\equiv D\{y' + v(x)\,y\} + u(x)\,\{y' + v(x)\,y\}\\ &\equiv \{y'' + v(x)\,y' + v'(x)\,y\} + \{u(x)\,y' + u(x)\,v(x)\,y\}\\ &\equiv y'' + \{u(x) + v(x)\}\,y' + \{v'(x) + u(x)\,v(x)\}\,y,\end{aligned}$$

we have the relations

$$u(x) + v(x) = P_1(x),$$

$$v'(x) + u(x)\,v(x) = P_2(x)$$

to determine $u(x)$, $v(x)$. Eliminating $u(x)$, we obtain an equation for $v(x)$ in the form

$$v'(x) + \{P_1(x) - v(x)\}\,v(x) = P_2(x),$$

or, writing

$$v(x) \equiv v,$$

$$\frac{dv}{dx} + P_1(x)\,v - v^2 = P_2(x).$$

This equation, known as a Riccati equation, is discussed in textbooks devoted to differential equations. The immediate need is any one solution, and the success of our attempts to factorize the operator will depend on our ability to find such a solution for the particular functions $P_1(x)$, $P_2(x)$ of the given equation.

In some problems the differential equation for $u(x)$ is more amenable, and it should be obtained if a value for $v(x)$ remains elusive. Since

$$v(x) = P_1(x) - u(x),$$

the equation is

$$P_1'(x) - u'(x) + u(x)\{P_1(x) - u(x)\} = P_2(x),$$

or, writing

$$u(x) \equiv u,$$

$$u' - P_1(x)\,u + u^2 = P_1'(x) - P_2(x),$$

another Riccati equation.

It should be remembered carefully that the order of the two factors $D + u(x)$, $D + v(x)$ must be preserved; they are NOT interchangeable.

ILLUSTRATION 6. *To solve the equation*

$$y'' + (\cot x - 2)\,y' - (\operatorname{cosec}^2 x + 2\cot x)\,y = e^x.$$

Suppose that the left-hand side is

$$\begin{aligned}(D+u)(D+v)\,y &\equiv D(y' + vy) + u(y' + vy)\\ &\equiv y'' + vy' + v'y + uy' + uvy\\ &\equiv y'' + (u+v)\,y' + (v' + uv)\,y.\end{aligned}$$

Then

$$u + v = \cot x - 2,$$

$$v' + uv = -\operatorname{cosec}^2 x - 2\cot x,$$

so that

$$v' + v(\cot x - 2 - v) = -\operatorname{cosec}^2 x - 2\cot x,$$

or

$$v' + v(\cot x - 2) - v^2 + (\operatorname{cosec}^2 x + 2\cot x) = 0.$$

By inspection (and this is a point of real difficulty for this method), a solution is

$$v = \cot x,$$

and it follows at once that $u = -2.$

Hence the equation is

$$(D-2)(D + \cot x)\,y = e^x.$$

Writing

$$(D + \cot x)\,y \equiv z,$$

we have

$$\frac{dz}{dx} - 2z = e^x,$$

and so, solving this first-order linear equation,

$$z = Ae^{2x} - e^x,$$

where A is an arbitrary constant. Hence

$$y' + y \cot x = Ae^{2x} - e^x,$$

or*
$$y' \sin x + y \cos x = Ae^{2x} \sin x - e^x \sin x,$$

so that
$$y \sin x = A \int e^{2x} \sin x \, dx + B - \int e^x \sin x \, dx,$$

where B is an arbitrary constant. Thus

$$y \sin x = \tfrac{1}{5}Ae^{2x}(2 \sin x - \cos x) + B - \tfrac{1}{2}e^x(\sin x - \cos x).$$

The difficulty in 'spotting' $v(x)$ suggests that we might have tried instead the equation for $u(x)$:

Since
$$v = \cot x - 2 - u,$$

the equation is

$$\{-\operatorname{cosec}^2 x - u'\} + u\{\cot x - 2 - u\} = -\operatorname{cosec}^2 x - 2 \cot x,$$

or
$$u' - u(\cot x - 2) + u^2 = 2 \cot x.$$

A glance at the coefficients of $\cot x$ suggests that we might consider the solution

$$u = -2,$$

and this is seen to be satisfactory. Then

$$v = \cot x,$$

and the solution proceeds as before.

REVISION EXAMPLES XVI

Solve the differential equations:

1. $\dfrac{dy}{dx} + 2y \tan x = \cos^5 x.$

2. $x(x+1)\dfrac{dy}{dx} - (x+2)\,y = x^3(x-3).$

3. $(x^2-1)\dfrac{dy}{dx} + y = (x^2-1)^{\frac{1}{2}}.$

4. $\dfrac{dy}{dx} + y \log x = e^{-x \log x}.$

* The equation is a linear equation of the type discussed on p. 19. The integrating factor is $\sin x$.

5. $x\dfrac{dy}{dx} + (x+3)\,y = \sinh x.$

6. $(1+x)(1+2x)\dfrac{dy}{dx} + y = (1+x)^2(3+2x).$

7. $x\log x\dfrac{dy}{dx} + y = \log x.$

8. $\dfrac{dy}{dx} + y\cot x = \sin x.$

9. $(x+1)\dfrac{dy}{dx} - 3y = (x+1)^5$, where $y = \frac{3}{2}$ when $x = 0$.

10. $\cot x\dfrac{dy}{dx} - y = \cot x \sin^4 x.$

11. $(x^2+3x+2)\dfrac{dy}{dx} + xy = x(x+1).$

12. $(4-\cos^2 x)\dfrac{dy}{dx} + 4y\sin x = \sin x(2+\cos x).$

13. $2x\dfrac{dy}{dx} - (1+2x\cot x)\,y = x^{\frac{3}{2}}\sin x.$

14. $\dfrac{dy}{dx} + \dfrac{x^2-x-1}{x-1}\,y = x(x-1).$

15. $\dfrac{dy}{dx} + y\sec x = \tan x.$

16. $(x^2-1)\dfrac{dy}{dx} + 2(x+2)\,y = 2(x+1).$

17. $x\dfrac{dy}{dx} - y = x.$

18. $\dfrac{dy}{dx} + \dfrac{y}{x} = \sin x$, where $y = 0$ when $x = \pi$.

19. $\cos x\dfrac{dy}{dx} + (\cos x + \sin x)\,y = 2 + \sin 2x.$

20. $x(x-1)\dfrac{dy}{dx} + (x-2)\,y = x^2.$

21. $(1-x^2)\,y'' - xy' = 2.$

22. $xy'' + y' + 1 = x.$

23. Solve the equation

$$(2x+x^2)\,y'' - 2(1+x)\,y' + 2y = 0,$$

given that $y = x^2$ is a solution.

24. Solve the equation

$$(1+x^2)^2\,y'' - 4x(1+x^2)\,y' - (1-8x^2-x^4)\,y = 2(1+x^2)^3\cos x,$$

given that one solution of the complementary equation is

$$(1+x^2)\sin x.$$

25. Show that the general solution of the equation

$$x^2y'' - 2xy' + 2y = 0$$

is a polynomial, and solve the equation

$$x^2y'' - 2xy' + 2y = x^3 \cos x.$$

26. Solve completely the equation

$$2x(1-x)^{a+1}\frac{d}{dx}\left\{2x(1-x)^{-a}\frac{dy}{dx}\right\} = \{1-(1+2a)\,x\}\,y,$$

given that there is a particular integral of the form $y = x^n$.

Examine the case $a = -1$.

27. Solve the equation

$$(2x^2-1)\,y'' - (4x^2+4x-2)\,y' + 8xy = 4x,$$

given that the complementary equation has a solution of the form $y = e^{ax}$.

28. Solve the differential equation

$$xy'' + (x-2)\,y' - 2y = x^3.$$

29. Solve, in as simple a form as you can, the differential equation

$$y'' + (x^2 - 2x^{-1})\,y' - 3xy = x^4,$$

given that the complementary equation has a solution which is a power of x.

30. Solve the differential equation

$$y'' + (1 + 2x^{-1}\cot x - 2x^{-2})\,y = x \cos x,$$

given that $x^{-1}\sin x$ is a solution of the complementary equation.

Prove also that, if $y = 1$ and $y' = 1$ when $x = 0$, then $y = \pi$ when $x = \pi$.

31. Solve the differential equation

$$(x \sin x + \cos x)\,y'' - xy' \cos x + y \cos x = \sin x(x \sin x + \cos x)^2,$$

given that x and $\cos x$ are solutions of the complementary equation.

32. Use the substitution $z = xy$ to solve the equation

$$xy'' + 2y' + a^2xy = 0,$$

where $y = 0$ and $y' = -a$ when $x = \pi/a$.

33. Explain what is meant by an integrating factor of the equation

$$P\,dx + Q\,dy = 0.$$

Show that, if P/Q is a function of y/x, then

$$(xP + yQ)^{-1}$$

is an integrating factor, giving the solution

$$\int \frac{d(y/x)}{y/x + P/Q} + \log x = C.$$

Solve the equation

$$(x^2 + y^2)\,dx - xy\,dy = 0.$$

34. Show that the equation

$$x^2(1 + x^2)\,y'' - 2y = 2x^4$$

possesses certain solutions of the form

$$y = Ax^{-1} + Bx + nx^2,$$

where A is an arbitrary constant, B a definite function of A, and n a definite number.

Find the general solution.

35. By putting $y = z \operatorname{cosec} x$, or otherwise, solve the differential equation

$$xy'' \sin x + 2y'(\sin x + x \cos x) + (2 \cos x - x \sin x)\,y = 3x.$$

36. Solve the differential equation

$$x^2y'' + xy' - 9y = 7x^4,$$

of which x^4 and $x^4 + x^3$ are particular solutions.

37. Find the relation between P and Q if the equation

$$y'' + Py' + Qy = 0$$

has two non-zero solutions one of which is the square of the other.

Show that this condition is satisfied for the equation

$$xy'' - (3x^2 + 1)\,y' + 2x^3y = 0,$$

and hence obtain the complete solution.

38. By factorizing the operator, or otherwise, solve the differential equation

$$y'' - 4xy' + (4x^2 - 2)\,y = 2x^3 - 3x.$$

39. By factorizing the operator, or otherwise, solve the differential equation

$$y'' - 2(n - ax^{-1})y' + (n^2 - 2nax^{-1})y = e^{nx},$$

where n, a are constants.

Examine the solution in the special cases $a = \pm \frac{1}{2}$.

40. By factorizing the operator in the form

$$\left(x\frac{d}{dx} + P\right)\left(x\frac{d}{dx} + Q\right),$$

or otherwise, solve the differential equation

$$x^2y'' + (x^2 + 4x)y' + 2(x+1)y = 4x^2(x+3).$$

41. If $y'' + Q(x)y' + R(x)y \equiv \left(\frac{d}{dx} - u(x)\right)\left(\frac{d}{dx} - v(x)\right)y,$

find a first-order differential equation, not involving v, satisfied by u.

Apply this to the equation

$$y'' - y' \tan x - 2y/(1 + \sin x) = 0,$$

and, using the substitution $u \cos x = z$, or otherwise, find a solution for u and hence solve completely the given equation.

42. The equation

$$\frac{d^2y}{dx^2} + P(x)\frac{dy}{dx} + Q(x)y = 0$$

has solutions $\cos x$ and $\tan x$. Find the general solution of the equation

$$\frac{d^2y}{dx^2} + P(x)\frac{dy}{dx} + Q(x)y = \frac{\cos x}{1 + \sin^2 x}.$$

43. Show that a necessary and sufficient condition for the expression

$$P(x)y'' + Q(x)y' + R(x)y$$

to be expressible in the form

$$\frac{d}{dx}\{L(x)y' + M(x)y\},$$

is that

$$P''(x) - Q'(x) + R(x) \equiv 0.$$

Solve completely the differential equation

$$x(1+x)y'' - \{n + (n-2)x\}y' - ny = x^{n+1}.$$

CHAPTER XXI

THE LINEAR DIFFERENTIAL EQUATION WITH CONSTANT COEFFICIENTS

The differential equations which we shall discuss, now and in Chapter XXII, are important in many branches of mathematics and physics. The form of the equations is obtained from the general linear equation (p. 12) by giving $P_0(x), P_1(x), \ldots$ constant values. The treatment adopted varies in some respects from current teaching practice, especially in the details of the two methods recommended for calculating 'particular integrals'.

When equations occur as often as these, their solution should conform to at least two principles: (i) it should follow a standard drill, (ii) that drill should rest on a simple logical foundation from which it arises naturally. The methods to be given seek to dispel the uncertainty which many beginners seem to feel about solutions by 'guessing the form of the answer', and also to avoid mechanical calculations involving operators like $(1-D)^{-1}$ which are often used without any appreciation of the underlying theory.

There is nothing essentially new in what follows, but the combination of 'guessing' with 'operation by *polynomials only* in D' gives a simple drill in which each step almost carries its own justification; and the later method based on integrals of the form

$$e^{-ax}\int Q(x)\,e^{ax}\,dx$$

leads to a 'calculus' which enables all particular integrals to be found in theory and all the usual ones in practice.

But first we must deal with the complementary function.

1. The linear operator for constant coefficients. The given equation is

$$a_0\frac{d^ny}{dx^n}+a_1\frac{d^{n-1}y}{dx^{n-1}}+\ldots+a_{n-1}\frac{dy}{dx}+a_ny=Q(x),$$

or $$L(D)\,y\equiv(a_0D^n+a_1D^{n-1}+\ldots+a_{n-1}D+a_n)\,y=Q(x),$$

where the coefficients $a_0, a_1, \ldots, a_{n-1}, a_n$ are constants.

The algebraic equation

$$L(p)\equiv a_0p^n+a_1p^{n-1}+\ldots+a_{n-1}p+a_n=0$$

is called the AUXILIARY EQUATION of the given equation and its roots

$$p_1, \quad p_2, \quad \ldots, \quad p_n,$$

not necessarily all different, the AUXILIARY ROOTS; thus

$$L(p) \equiv a_0(p-p_1)(p-p_2)\ldots(p-p_n).$$

For example, the auxiliary equation of

$$\frac{d^3y}{dx^3} - \frac{d^2y}{dx^2} - 8\frac{dy}{dx} + 12y = 5\sin 7x$$

is
$$p^3 - p^2 - 8p + 12 = 0,$$

or
$$(p+3)(p-2)^2 = 0,$$

and the auxiliary roots are -3, 2, 2.

It is familiar from the theory of equations that the auxiliary roots are connected with the coefficients in the auxiliary equation by means of the formulae

$$\begin{aligned} p_1 + p_2 + p_3 + \ldots &= -a_1/a_0, \\ p_1p_2 + p_1p_3 + \ldots + p_2p_3 + \ldots &= a_2/a_0, \\ p_1p_2p_3 + p_1p_2p_4 + \ldots + p_1p_3p_4 + \ldots &= -a_3/a_0, \end{aligned}$$

and so on. Now, by continued operations,

$$(D-p_1)(D-p_2) = D^2 - (p_1+p_2)D + p_1p_2,$$

$$\begin{aligned} &(D-p_1)(D-p_2)(D-p_3) \\ &\quad = D^3 - (p_1+p_2+p_3)D^2 + (p_1p_2+p_1p_3+p_2p_3)D - p_1p_2p_3, \end{aligned}$$

and so on. From the 'product' of n such operators, we obtain, with the help of the algebraic identities just quoted, the relation

$$\begin{aligned} &(D-p_1)(D-p_2)\ldots(D-p_n) \\ &\quad \equiv D^n - (-a_1/a_0)D^{n-1} + (a_2/a_0)D^{n-2} - (-a_3/a_0)D^{n-3} + \ldots \\ &\quad \equiv \frac{1}{a_0}\{a_0D^n + a_1D^{n-1} + a_2D^{n-2} + a_3D^{n-3} + \ldots\}. \end{aligned}$$

Thus the operator may be expressed in the 'factorized' form

$$L(D)\,y \equiv a_0(D-p_1)(D-p_2)\ldots(D-p_n)\,y,$$

where it is important to observe that, in this particular case, *the factors $D-p_1, D-p_2, \ldots, D-p_n$ may be taken in any order without affecting the operator.*

For example,

$$\begin{aligned}(D^3-D^2-8D+12)\,y &\equiv (D+3)\,(D-2)^2\,y\\ &\equiv (D-2)^2\,(D+3)\,y\\ &\equiv (D-2)\,(D+3)\,(D-2)\,y.\end{aligned}$$

2. The complementary function.* In accordance with the work of the preceding paragraph, the complementary equation is

$$a_0(D-p_1)\,(D-p_2)\ldots(D-p_n)\,y=0.$$

Suppose that any one root of the auxiliary equation is denoted by the letter p, and that it is a k-fold root. Factorizing the operator in such a way that the factors corresponding to p appear last, we obtain the equation in the form

$$a_0(D-p_1)\ldots(D-p_{n-k})\,(D-p)^k\,y=0.$$

If we are able to obtain a function y satisfying the relation

$$(D-p)^k\,y=0,$$

we shall also have as a consequence the relation

$$a_0(D-p_1)\ldots(D-p_{n-k})\,\{(D-p)^k\,y\}=0,$$

and such a function y will thus be a solution of the complementary equation. We therefore begin by considering the equation

$$(D-p)^k\,y=0.$$

When $k=1$, this is an ordinary linear equation

$$(D-p)\,y=0$$

whose solution is $y=ae^{px}$. This suggests the method of variation of parameters, using the substitution

$$y=ze^{px}.$$

* This paragraph should be read quickly at first in order to reach the subsequent Illustrations as soon as possible. These should be studied very carefully so as to absorb the 'drill'.

Observe, as a *Lemma*, that, if u is any function of x, then

$$\begin{aligned}(D-p)(ue^{px}) &= D(ue^{px}) - pue^{px} \\ &= \{u'e^{px} + pue^{px}\} - pue^{px} \\ &= u'e^{px},\end{aligned}$$

so that $$(D-p)(ue^{px}) = u'e^{px}.$$

If, therefore, we give to u the values $z, z', z'', \ldots$ in succession, we obtain the formulae

$$(D-p)(ze^{px}) = z'e^{px},$$

$$(D-p)^2(ze^{px}) = (D-p)(z'e^{px}) = z''e^{px},$$

$$(D-p)^3(ze^{px}) = (D-p)(z''e^{px}) = z'''e^{px},$$

and so on. Hence $$(D-p)^k(ze^{px}) = z^{(k)}e^{px}.$$

The equation

$$(D-p)^k y = 0,$$

or

$$(D-p)^k(ze^{px}) = 0,$$

is therefore

$$z^{(k)}e^{px} = 0,$$

or

$$z^{(k)} = 0,$$

since e^{px} is not zero. Hence *z is an arbitrary polynomial in x of degree $k-1$, and the corresponding contribution to the complementary function is*

$$y = (A_1 + A_2 x + \ldots + A_k x^{k-1})e^{px},$$

where $A_1, A_2, \ldots, A_k$ are arbitrary constants.

Repeating this process for the other auxiliary roots, we reach the following RULE FOR THE FORMATION OF THE COMPLEMENTARY FUNCTION:

If the auxiliary equation has roots p (repeated α times), q (repeated β times), r (repeated γ times), …, where

$$\alpha + \beta + \gamma + \ldots = n,$$

then the complementary function, with its n arbitrary constants, is

$$\begin{aligned}&(A_1 + A_2 x + \ldots + A_\alpha x^{\alpha-1})e^{px} \\ &\quad + (B_1 + B_2 x + \ldots + B_\beta x^{\beta-1})e^{qx} \\ &\quad\quad + (C_1 + C_2 x + \ldots + C_\gamma x^{\gamma-1})e^{rx} \\ &\quad + \ldots.\end{aligned}$$

We have proved (pp. 15–16) that the n functions of the type $x^i e^{p_j x}$ occurring in this expression are linearly independent in the two cases (i) $\alpha = \beta = \gamma = \ldots = 1$, (ii) $\alpha = n$, $\beta = \gamma = \ldots = 0$. The other cases are more difficult, and we propose to accept the fact of their independence without further proof.

ILLUSTRATION 1. *To find the complementary function of the equation*

$$\frac{d^2y}{dx^2} + 4\frac{dy}{dx} - 5y = 1 + x^3.$$

The auxiliary equation is

$$p^2 + 4p - 5 = 0,$$

or
$$(p-1)(p+5) = 0.$$

Hence
$$p = 1, -5,$$

and the complementary function is

$$Ae^x + Be^{-5x}.$$

ILLUSTRATION 2. *To find the complementary function of the equation*

$$\frac{d^2y}{dx^2} + 9\frac{dy}{dx} + 20y = e^{7x}.$$

The auxiliary equation is

$$p^2 + 9p + 20 = 0,$$

or
$$(p+4)(p+5) = 0.$$

Hence
$$p = -4, -5,$$

and the complementary function is

$$Ae^{-4x} + Be^{-5x}.$$

ILLUSTRATION 3. *To find the complementary function of the equation*

$$\frac{d^3y}{dx^3} - 6\frac{d^2y}{dx^2} + 11\frac{dy}{dx} - 6y = x^2.$$

The auxiliary equation is

$$p^3 - 6p^2 + 11p - 6 = 0,$$

or
$$(p-1)(p-2)(p-3) = 0.$$

Hence $$p = 1, 2, 3,$$

and the complementary function is

$$Ae^{x} + Be^{2x} + Ce^{3x}.$$

ILLUSTRATION 4. *To find the complementary function of the equation*

$$\frac{d^4y}{dx^4} - 24\frac{d^2y}{dx^2} + 64\frac{dy}{dx} - 48y = 5\sin 3x.$$

The auxiliary equation is

$$p^4 - 24p^2 + 64p - 48 = 0,$$

or $$(p+6)(p-2)^3 = 0.$$

Hence $$p = -6, 2, 2, 2,$$

and the complementary function is

$$Ae^{-6x} + (B + Cx + Dx^2)\, e^{2x}.$$

ILLUSTRATION 5. *To find the complementary function of the equation*

$$\frac{d^4y}{dx^4} - 8\frac{d^2y}{dx^2} + 16y = x^2e^{x}.$$

The auxiliary equation is

$$p^4 - 8p^2 + 16 = 0,$$

or $$(p-2)^2(p+2)^2 = 0.$$

Hence $$p = 2, 2, -2, -2,$$

and the complementary function is

$$(A + Bx)\, e^{2x} + (C + Dx)\, e^{-2x}.$$

3. The complementary function; complex roots. The preceding work is true whether the roots $p_1, p_2, \ldots, p_n$ are real or complex, but alternative forms of expression prove more convenient for the latter.

It is assumed that the coefficients $a_0, a_1, \ldots, a_n$ in the differential equation are all real; then complex roots of the auxiliary equation occur in conjugate pairs. Suppose that p_1, p_2 constitute such a pair, of the form $\alpha + i\beta$, $\alpha - i\beta$ respectively, repeated k times. The corresponding contribution to the complementary function is

$$(A_1 + A_2x + \ldots + A_kx^{k-1})\, e^{(\alpha+i\beta)x} + (B_1 + B_2x + \ldots + B_kx^{k-1})\, e^{(\alpha-i\beta)x}.$$

Consider the terms in a typical power of x, say x^{j-1}. These are

$$A_j x^{j-1} e^{(\alpha+i\beta)x} + B_j x^{j-1} e^{(\alpha-i\beta)x},$$

or
$$x^{j-1} e^{\alpha x} \{A_j e^{i\beta x} + B_j e^{-i\beta x}\},$$

or (Volume II, p. 179)

$$x^{j-1} e^{\alpha x} \{A_j(\cos\beta x + i\sin\beta x) + B_j(\cos\beta x - i\sin\beta x)\},$$

or
$$x^{j-1} e^{\alpha x} \{(A_j + B_j)\cos\beta x + (iA_j - iB_j)\sin\beta x\}.$$

The constants A_j, B_j are complex, and may be replaced by the complex numbers P_j, Q_j defined by the relations

$$A_j + B_j = P_j,$$
$$iA_j - iB_j = Q_j.$$

Since the constants A_j, B_j can, in the first instance, have any values, so also can P_j, Q_j.

The contribution from the terms in x^{j-1} then assumes the form

$$x^{j-1} e^{\alpha x} \{P_j \cos\beta x + Q_j \sin\beta x\},$$

and so *the total contribution involving* $p_1 = \alpha + i\beta$ *and* $p_2 = \alpha - i\beta$ *is*

$$(P_1 + P_2 x + \dots + P_k x^{k-1})\, e^{\alpha x}\cos\beta x$$
$$+ (Q_1 + Q_2 x + \dots + Q_k x^{k-1})\, e^{\alpha x}\sin\beta x.$$

When $k=1$ (which the reader is most likely to need in practice), the contribution is
$$e^{\alpha x}(A\cos\beta x + B\sin\beta x).$$

When $k=1$ and $\alpha=0$, we have the familiar case of 'harmonic motion', with solution
$$A\cos\beta x + B\sin\beta x.$$

The 'harmonic equation' itself is of the form

$$\frac{d^2y}{dx^2} + n^2 y = 0,$$

and the solution
$$y = A\cos nx + B\sin nx$$

may thus be written down at once.

It can be verified easily that the general solution of the equation

$$\frac{d^2y}{dx^2} - m^2 y = 0$$

may likewise be expressed in the form

$$y = A\cosh mx + B\sinh mx.$$

ILLUSTRATION 6. *To solve the equation*

$$\frac{d^4y}{dx^4} + 6\frac{d^2y}{dx^2} + 9y = 0.$$

The auxiliary equation is

$$p^4 + 6p^2 + 9 = 0,$$

or

$$(p^2 + 3)^2 = 0,$$

so that

$$p = i\sqrt{3},\ i\sqrt{3},\ -i\sqrt{3},\ -i\sqrt{3}.$$

The solution is therefore

$$y = (A + Bx)\cos x\sqrt{3} + (C + Dx)\sin x\sqrt{3}.$$

ILLUSTRATION 7. *To solve the equation*

$$\frac{d^2y}{dx^2} + 4\frac{dy}{dx} + 13y = 0.$$

The auxiliary equation is

$$p^2 + 4p + 13 = 0,$$

so that

$$p = -2 + 3i,\ -2 - 3i.$$

The solution is therefore

$$y = e^{-2x}(A\cos 3x + B\sin 3x).$$

EXAMPLES I

Solve the following linear differential equations:

1. $y'' - 3y' + 2y = 0$.
2. $2y'' + 5y' + 2y = 0$.
3. $y'' + 9y = 0$.
4. $y^{\text{iv}} + 2y'' + y = 0$.
5. $y'' + 8y' + 25y = 0$.
6. $y'' + y' - y = 0$.
7. $y'' - 2y' + 17y = 0$.
8. $y''' - 3y'' + 3y' - y = 0$.
9. $y''' + 6y'' + 12y' + 8y = 0$.
10. $y'' + 10y' + 26y = 0$.

Further examples, if required, may be obtained by equating to zero the left-hand sides of nos. 1–19 and 21–49 of Revision Examples XVII, pp. 55–57.

4. A particular integral; rule for 'normal' cases. To find a particular integral of the equation

$$L(D)\,y \equiv (a_0 D^n + a_1 D^{n-1} + \ldots + a_n)\,y = Q(x),$$

we must give closer attention than hitherto to the form of the function $Q(x)$. It is found by experience that, in most of the examples likely to be met at present, $Q(x)$ is a sum of terms consisting of the product of one or more of the functions: polynomial, sine or cosine, exponential. Remembering that sines and cosines are essentially exponential, we may take $Q(x)$ as a sum of terms like

$$f(x)\,e^{bx},$$

where $f(x)$ is a *polynomial* and b a complex number, possibly zero. It is a simple matter, too, to prove that *a particular integral corresponding to a sum of terms may be found by adding particular integrals corresponding to the individual components of the sum.*

Suppose, then, that $f(x)$ is a polynomial of degree m, and that b is a k-fold root of the auxiliary equation $L(p) = 0$. We begin with a lemma designed to remove the exponential e^{bx} from the calculations.

LEMMA. To prove that, *if u is any function of x, then*

$$L(D)\,\{ue^{bx}\} = e^{bx}\,L(D+b)\,u,$$

where $L(D+b)$, found by replacing D by $D+b$ in $L(D)$, is given by the relation

$$\begin{aligned} L(D+b) &\equiv a_0(D+b)^n + a_1(D+b)^{n-1} + \ldots + a_n \\ &\equiv a_0(D+b-p_1)\,(D+b-p_2)\ldots(D+b-p_n). \end{aligned}$$

The proof is by induction; for

$$\begin{aligned} D(ue^{bx}) &= u'e^{bx} + ube^{bx} \\ &= e^{bx}\left(\frac{du}{dx} + bu\right) \\ &= e^{bx}\,(D+b)\,u. \end{aligned}$$

Hence
$$D^2(ue^{bx}) = D\{e^{bx}[(D+b)u]\}$$
$$= e^{bx}(D+b)[(D+b)u]$$
$$= e^{bx}(D+b)^2 u,$$
$$D^3(ue^{bx}) = D\{e^{bx}[(D+b)^2 u]\}$$
$$= e^{bx}(D+b)[(D+b)^2 u]$$
$$= e^{bx}(D+b)^3 u,$$

and so on; and the result follows by simple substitution in the polynomial operator $L(D)$.

NOTE. To evaluate $L(D+b)$, in practice, either of the forms

$$a_0(D+b)^n + a_1(D+b)^{n-1} + \dots + a_n$$

or

$$a_0(D+b-p_1)(D+b-p_2)\dots(D+b-p_n)$$

may be used. The latter has advantages if the auxiliary equation has been factorized previously.

We now proceed to the first step in the calculation of a particular integral:

To prove that *if, in the equation*

$$L(D)\,y = f(x)\,e^{bx},$$

the function y is written in the form

$$y = ze^{bx},$$

then z satisfies the differential equation

$$L(D+b)\,z = f(x).$$

The proof follows from the Lemma; for, by it,

$$L(D)\,ze^{bx} = e^{bx}\,L(D+b)\,z,$$

so that the equation becomes

$$e^{bx}\,L(D+b)\,z = f(x)\,e^{bx},$$

or

$$L(D+b)\,z = f(x).$$

ILLUSTRATION 8. *To reduce the equation*

$$\frac{d^2y}{dx^2} - 9\frac{dy}{dx} + 20y = (1+x^2)\,e^{4x}.$$

The equation is

$$(D-4)(D-5)y=(1+x^2)e^{4x}.$$

If

$$y=ze^{4x},$$

then, by the Lemma,

$$e^{4x}(D-4+4)(D-5+4)z=(1+x^2)e^{4x},$$

so that

$$D(D-1)z=1+x^2,$$

or

$$\frac{d^2z}{dx^2}-\frac{dz}{dx}=1+x^2.$$

ILLUSTRATION 9. *To reduce the equation*

$$\frac{d^2y}{dx^2}+4\frac{dx}{dy}+4y=7e^{-2x}.$$

The equation is

$$(D+2)^2y=7e^{-2x}.$$

If

$$y=ze^{-2x},$$

then, by the Lemma,

$$e^{-2x}(D+2-2)^2z=7e^{-2x},$$

so that

$$D^2z=7,$$

or

$$\frac{d^2z}{dx^2}=7.$$

In this simple case, we can find a particular function z at once, namely,

$$z=\tfrac{7}{2}x^2.$$

Thus a particular solution of the given equation is

$$y=\tfrac{7}{2}x^2e^{-2x}.$$

The next step is to devise a technique for obtaining a particular integral of the simplified equation

$$L(D+b)z=f(x)$$

when $f(x)$ is a given *polynomial*. To do this, we enunciate, without proof, a general rule, but emphasize that, as we have already remarked, the actual process of calculation from that rule will of itself form the justification in any particular example. We give a number of illustrations to show how the rule may be applied in such a way as to keep the working as economical as possible. (There is danger of floundering among linear algebraic equations unless the treatment is kept systematic.)

THE RULE* TO DETERMINE z. *If z satisfies the differential equation*

$$L(D+b)\,z=f(x),$$

where $f(x)$ is a polynomial of degree m and b is a k-fold root of the auxiliary equation $L(p)=0$, then there exists a particular solution

$$z=x^k(U_1+U_2x+\ldots+U_{m+1}x^m)$$

for suitable values of the constants $U_1, U_2, \ldots, U_{m+1}$.

The values of $U_1, U_2, \ldots, U_{m+1}$ are to be found by substituting this expression in the function $L(D+b)\,z$ and identifying the result with the given polynomial $f(x)$.

The Illustrations which follow are of the standard of difficulty which the student may expect at this stage, and show how the coefficients $U_1, U_2, \ldots, U_{m+1}$ are obtained in practice.

ILLUSTRATION 10. *To find a particular integral of the equation*

$$\frac{d^2y}{dx^2}-5\frac{dy}{dx}+6y=(7x+9)\,e^{-2x}.$$

If $$y=ze^{-2x},$$
the equation for z, namely,

$$L(D+b)\,z=f(x),$$

gives $$\{(D-2)^2-5(D-2)+6\}\,z=7x+9,$$

or $$(D^2-9D+20)\,z=7x+9.$$

The polynomial $7x+9$ is of degree 1, and -2 is a '0-ple' root of the auxiliary equation $p^2-5p+6=0$. Hence we seek a particular solution (with $m=1$, $k=0$)

$$z=x^0(U_1+U_2x)$$
$$=U_1+U_2x.$$

* When $k=m=0$, corresponding to the equation

$$L(D)\,y=ae^{bx}\quad (a \text{ constant}),$$

a particular solution $$y=\frac{ae^{bx}}{L(b)}\quad (L(b)\neq 0)$$

can be written down at once. See the Corollary on pp. 43–4.

Since then $$z' = U_2,$$
$$z'' = 0,$$
we require the identity
$$-9U_2 + 20(U_1 + U_2 x) \equiv 7x + 9,$$
or
$$20U_2 = 7,$$
$$20U_1 - 9U_2 = 9.$$
Thus
$$U_2 = \tfrac{7}{20}, \quad U_1 = \tfrac{243}{400}.$$
Hence a particular integral is
$$y = \left(\tfrac{243}{400} + \tfrac{7}{20}x\right) e^{-2x}.$$

ILLUSTRATION 11. *To find a particular integral of the equation*
$$\frac{d^2y}{dx^2} - 3\frac{dy}{dx} + 2y = 7e^{bx}.$$

If
$$y = ze^{bx},$$
then the equation $L(D+b)z = f(x)$ gives
$$\{(D+b)^2 - 3(D+b) + 2\} z = 7,$$
or
$$\{D^2 + (2b-3)D + (b^2 - 3b + 2)\} z = 7.$$

In order to apply the rule, we have to find the value of k; we know that $m = 0$, since 7 is a pure constant.

(i) If $b \neq 1$, $b \neq 2$, then $k = 0$, and so z is of the form
$$z = U_1,$$
and substitution gives at once the relation
$$(b^2 - 3b + 2) U_1 = 7,$$
or
$$U_1 = \frac{7}{b^2 - 3b + 2};$$
thus
$$y = \frac{7e^{bx}}{b^2 - 3b + 2}.$$

COROLLARY. Identical treatment leads in the general case to the SUBSIDIARY RULE, which is convenient in practice:

A particular solution of the equation
$$L(D) y = ae^{bx},$$

where b is NOT *a root of the auxiliary equation, is*

$$y=\frac{ae^{bx}}{L(b)}.$$

This rule is very special, but worth remembering because of its simplicity.

(ii) (α) If $b=1$, then $k=1$, and so z is of the form

$$z=xU_1,$$

and substitution in the equation

$$(D^2-D)\,xU_1=7$$

gives the relation $$-U_1=7,$$

so that $$z=-7x$$

and $$y=-7xe^x.$$

(β) If $b=2$, then $k=1$, and so z is of the form

$$z=xU_1$$

(for fresh U_1), and substitution in the equation

$$(D^2+D)\,xU_1=7$$

gives the relation $$U_1=7,$$

so that $$z=7x$$

and $$y=7xe^{2x}.$$

ILLUSTRATION 12. *To find a particular integral of the equation*

$$\frac{d^2y}{dx^2}-3\frac{dy}{dx}+2y=2x^2+4x+13.$$

(There is no exponential on the right-hand side, so the preliminary step of removing it does not arise. Also the polynomial $2x^2+4x+13$ may, for the moment, be regarded in the form $(2x^2+4x+13)\,e^{0.x}$, with b interpreted as zero. But 0 is not a root of the auxiliary equation, so that $k=0$ in the rule.)

Assume the existence of a particular integral in the form

$$y=U+Vx+Wx^2.$$

Then $$y'=V+2Wx,$$

$$y''=2W.$$

This value of y is a solution if (tabulating the summation in a form which explains itself)

$$\left\{\begin{array}{r|l} 2y & 2U + 2Vx + 2Wx^2 \\ -3y' & -3V - 6Wx \\ y'' & 2W \\ \hline \equiv f(x) & 13 \quad + 4x \quad + 2x^2 \end{array}\right. .$$

Equate coefficients of x^2, x and the constant in turn. Hence

$$2W = 2, \quad \text{or} \quad W = 1;$$

$$-6W + 2V = 4, \quad \text{or} \quad V = 5;$$

$$2W - 3V + 2U = 13, \quad \text{or} \quad U = 13.$$

Hence a particular integral is

$$13 + 5x + x^2.$$

ILLUSTRATION 13. *To find a particular integral of the equation*

$$\frac{d^2y}{dx^2} - 3\frac{dy}{dx} + 2y = 3x^2e^{2x}.$$

If
$$y = ze^{2x},$$

then the equation $L(D+b)z = f(x)$ gives

$$\{(D+2)^2 - 3(D+2) + 2\}z = 3x^2,$$

or
$$(D^2 + D)z = 3x^2.$$

(This equation may be integrated one stage at once, or the rule may be applied directly. We choose the latter alternative, which will be found to have definite advantages.)

Assume the existence of a particular integral in the form (with $k = 1$)

$$z = x(U + Vx + Wx^2)$$

$$= Ux + Vx^2 + Wx^3.$$

Then
$$z' = U + 2Vx + 3Wx^2,$$

$$z'' = 2V + 6Wx.$$

This value of z is therefore a solution if

$$\begin{array}{r|l} \left\{\begin{array}{l} z' \\ z'' \end{array}\right. & \begin{array}{l} U+2Vx+3Wx^2 \\ 2V+6Wx \end{array} \\ \hline \equiv f(x) & \qquad\qquad 3x^2. \end{array}$$

Equating coefficients of x^2, x, and the constant in turn, we have the equations

$$3W=3, \quad \text{or} \quad W=1;$$

$$2V+6W=0, \quad \text{or} \quad V=-3;$$

$$U+2V=0, \quad \text{or} \quad U=6.$$

Hence a particular integral of the given equation is

$$x(6-3x+x^2)\,e^{2x}.$$

ILLUSTRATION 14. *To find a particular integral of the equation*

$$\frac{d^2y}{dx^2}-4\frac{dy}{dx}+13y=12xe^{2x}\sin 3x.$$

The solution of this equation is the imaginary part of the solution of the corresponding equation in which $\sin 3x$ is replaced by e^{3ix}. We thus consider the equation

$$\frac{d^2y}{dx^2}-4\frac{dy}{dx}+13y=12xe^{(2+3i)x}.$$

The auxiliary equation is

$$p^2-4p+13=0,$$

or

$$(p-2-3i)\,(p-2+3i)=0,$$

so that $2+3i$ is a simple root, and $k=1$.

If

$$y=ze^{(2+3i)x},$$

then the equation $L(D+b)=f(x)$ gives

$$\{(D-2-3i)+(2+3i)\}\,\{(D-2+3i)+(2+3i)\}\,z=12x,$$

or

$$(D^2+6iD)\,z=12x.$$

Assume the existence of a particular integral in the form

$$z=x(U+Vx)$$

$$=Ux+Vx^2$$

Then $$z' = U + 2Vx,$$

$$z'' = 2V.$$

This value of z is therefore a solution if

$$\left\{\begin{array}{r|l} 6iz' & 6iU + 12iVx \\ z'' & 2V \\ \hline \equiv f(x) & \quad 12x. \end{array}\right.$$

Equating coefficients of x and constant, we have

$$12iV = 12, \quad \text{or} \quad V = -i;$$

$$6iU + 2V = 0, \quad \text{or} \quad U = \tfrac{1}{3}.$$

Hence a particular integral of the given equation is

$$\mathscr{I}\{x(\tfrac{1}{3} - ix)\, e^{(2+3i)x}\}$$

$$= \mathscr{I}\{x(\tfrac{1}{3} - ix)\, e^{2x}(\cos 3x + i \sin 3x)\}$$

$$= xe^{2x}(\tfrac{1}{3}\sin 3x - x\cos 3x).$$

Illustration 15. *To solve the equation*

$$\frac{d^3y}{dx^3} + \frac{d^2y}{dx^2} - 5\frac{dy}{dx} + 3y = 96(x+1)\, e^x.$$

(This Illustration is included to show how the full solution may be set out in practice.)

The auxiliary equation is

$$p^3 + p^2 - 5p + 3 = 0,$$

or $$(p+3)(p-1)^2 = 0,$$

so that $$p = -3, 1, 1,$$

and the complementary function is

$$Ae^{-3x} + (B + Cx)\, e^x.$$

For the particular integral, write

$$y = ze^x.$$

Then (using the factorized form of the operator)

$$(\overline{D+3} + 1)(\overline{D-1} + 1)^2 z = 96(x+1),$$

or $$(D^3 + 4D^2)\, z = 96(x+1).$$

Assume the existence of a particular integral in the form

$$z = x^2(U + Vx)$$
$$= Ux^2 + Vx^3.$$

Then $$z' = 2Ux + 3Vx^2,$$
$$z'' = 2U + 6Vx,$$
$$z''' = 6V.$$

This value of z is therefore a solution if

$$\begin{array}{r|ll} \left\{\begin{array}{r} 4z'' \\ z''' \end{array}\right. & \begin{array}{l} 8U + 24Vx \\ 6V \end{array} & \\ \hline \equiv f(x) & 96 & +96x. \end{array}$$

Equating coefficients of x, constant, in turn, we have the equations

$$24V = 96, \quad \text{or} \quad V = 4;$$
$$8U + 24 = 96, \quad \text{or} \quad U = 9.$$

Hence the solution of the given equation is

$$y = Ae^{-3x} + (B + Cx)\, e^x + x^2 e^x\, (9 + 4x).$$

For Examples on the work of § 4, see Revision Examples XVII, nos. 1–49 (pp. 55–57). Where particular conditions are given, work out the general solution first.

5. Simultaneous differential equations. Without going into great detail, we give illustrations to indicate the process of solving two given linear differential equations with constant coefficients. The first method is of limited application, but we begin with it in order to emphasize the advantages of not turning blindly to the more routine procedure of the second until alternatives have been considered.

ILLUSTRATION 16. *To solve the simultaneous linear equations*

$$5\frac{d^2x}{dt^2} + 4x + 48y = 10e^t,$$

$$5\frac{d^2y}{dt^2} + 8x - 4y = 5t.$$

(The method which follows is applicable when the two equations involve only ONE of the operators d/dt, d^2/dt^2, d^3/dt^3, ...)

Multiply the second equation by λ and add to the first. Then

$$5\frac{d^2x}{dt^2}+5\lambda\frac{d^2y}{dt^2}+(4+8\lambda)\,x+(48-4\lambda)\,y=10e^t+5\lambda t.$$

Choose λ so that the coefficients of d^2x/dt^2 and d^2y/dt^2 are proportional to those of x and y; thus

$$\frac{5}{5\lambda}=\frac{4+8\lambda}{48-4\lambda},$$

or

$$\frac{1}{\lambda}=\frac{1+2\lambda}{12-\lambda}.$$

Hence $$2\lambda^2+\lambda=12-\lambda,$$

or $$\lambda^2+\lambda-6=0,$$

so that $$\lambda=2 \quad \text{or} \quad -3.$$

(i) Take $\lambda=2$; then

$$5\left(\frac{d^2x}{dt^2}+2\frac{d^2y}{dt^2}\right)+20(x+2y)=10e^t+10t.$$

Writing $$x+2y\equiv u,$$

we have $$\frac{d^2u}{dt^2}+4u=2e^t+2t.$$

Using methods with which the reader is now familiar, we have

$$u=A\cos 2t+B\sin 2t+\tfrac{2}{5}e^t+\tfrac{1}{2}t.$$

(ii) Take $\lambda=-3$; then

$$5\left(\frac{d^2x}{dt^2}-3\frac{d^2y}{dt^2}\right)-20(x-3y)=10e^t-15t.$$

Writing $$x-3y=v,$$

we have $$\frac{d^2v}{dt^2}-4v=2e^t-3t,$$

so that $$v=Pe^{2t}+Qe^{-2t}-\tfrac{2}{3}e^t+\tfrac{3}{4}t.$$

(Note, as a trivial point, that we have avoided the letters C, D for arbitrary constants so that no confusion may arise with the use of D as an operator.)

We have therefore reached the equations

$$x+2y=A\cos 2t+B\sin 2t+\tfrac{2}{5}e^t+\tfrac{1}{2}t,$$

$$x-3y=Pe^{2t}+Qe^{-2t}-\tfrac{2}{3}e^t+\tfrac{3}{4}t,$$

which, when solved for x, y, give the result

$$x=\tfrac{3}{5}A\cos 2t+\tfrac{3}{5}B\sin 2t+\tfrac{2}{5}Pe^{2t}+\tfrac{2}{5}Qe^{-2t}-\tfrac{2}{75}e^t+\tfrac{3}{5}t,$$

$$y=\tfrac{1}{5}A\cos 2t+\tfrac{1}{5}B\sin 2t-\tfrac{1}{5}Pe^{2t}-\tfrac{1}{5}Qe^{-2t}+\tfrac{16}{75}e^t-\tfrac{1}{20}t.$$

ILLUSTRATION 17. *To solve the simultaneous linear equations*

$$5\frac{d^2x}{dt^2}+4x+48y=10e^t,$$

$$5\frac{d^2y}{dt^2}+8x-\ \ 4y=5t.$$

(These are the equations solved in the preceding Illustration. The procedure is more general, and can be applied to most of the equations likely to be met.)

In terms of the operator D, the equations are

$$(5D^2+4)x+48y=10e^t,$$

$$8x+(5D^2-4)y=5t.$$

The method is very similar to that used for solving ordinary algebraic equations, save that we eliminate by operators and not by multiplication:

Eliminate y by operating on the first equation by $5D^2-4$ and the second by 48 (in this particular case a numerical multiplier only) and subtracting. Thus

$$\{(5D^2-4)(5D^2+4)-48.8\}x=10(5D^2-4)e^t-48.5t,$$

or

$$(25D^4-400)x=10e^t-240t,$$

or

$$(D^4-16)x=\tfrac{2}{5}e^t-\tfrac{48}{5}t.$$

Following the normal rules, we have the solution

$$x=A\cos 2t+B\sin 2t+Pe^{2t}+Qe^{-2t}-\tfrac{2}{75}e^t+\tfrac{3}{5}t.$$

Two courses are now possible, and we illustrate each in turn:

(i) Eliminate x by operating on the second equation by $(5D^2+4)$ and the first by 8 and subtracting. Thus

$$\{(5D^2+4)(5D^2-4)-8.48\}y=(5D^2+4)5t-8.10e^t,$$

or $$(25D^4 - 400)\,y = 20t - 80e^t,$$

or $$(D^4 - 16)\,y = \tfrac{4}{5}t - \tfrac{16}{5}e^t.$$

Hence $$y = A'\cos 2t + B'\sin 2t + P'e^{2t} + Q'e^{-2t} - \tfrac{1}{20}t + \tfrac{16}{75}e^t.$$

This solution involves *eight* arbitrary constants, and we cannot be sure without further investigation that they are independent. *It is necessary when using this method to check the proposed values in one of the given equations:*

Substituting in the first of the given equations, we obtain the relation

$$\begin{aligned} &5(-4A\cos 2t - 4B\sin 2t + 4Pe^{2t} + 4Qe^{-2t} - \tfrac{2}{75}e^t)\\ &+4(A\cos 2t + B\sin 2t + Pe^{2t} + Qe^{-2t} - \tfrac{2}{75}e^t + \tfrac{3}{5}t)\\ &+48(A'\cos 2t + B'\sin 2t + P'e^{2t} + Q'e^{-2t} + \tfrac{16}{75}e^t - \tfrac{1}{20}t) \equiv 10e^t. \end{aligned}$$

Equating the coefficients of $\cos 2t$, $\sin 2t$, e^{2t}, e^{-2t} in turn (and checking that the other terms cancel automatically) we have

$$-16A + 48A' = 0,$$

$$-16B + 48B' = 0,$$

$$24P + 48P' = 0,$$

$$24Q + 48Q' = 0,$$

so that $\quad A' = \tfrac{1}{3}A,\quad B' = \tfrac{1}{3}B,\quad P' = -\tfrac{1}{2}P,\quad Q' = -\tfrac{1}{2}Q.$

The solution is thus

$$x = A\cos 2t + B\sin 2t + Pe^{2t} + Qe^{-2t} - \tfrac{2}{75}e^t + \tfrac{3}{5}t,$$

$$y = \tfrac{1}{3}A\cos 2t + \tfrac{1}{3}B\sin 2t - \tfrac{1}{2}Pe^{2t} - \tfrac{1}{2}Qe^{-2t} + \tfrac{16}{75}e^t - \tfrac{1}{20}t.$$

(ii) The value of y may be obtained directly from the first of the given equations. (But this method, again, is limited in scope to equations for which such a solution is possible.) Thus, from the first of the given equations,

$$\begin{aligned} 48y = &-5(-4A\cos 2t - 4B\sin 2t + 4Pe^{2t} + 4Qe^{-2t} - \tfrac{2}{75}e^t)\\ &-4(A\cos 2t + B\sin 2t + Pe^{2t} + Qe^{-2t} - \tfrac{2}{75}e^t + \tfrac{3}{5}t)\\ &+10e^t, \end{aligned}$$

so that

$$y = \tfrac{1}{3}A\cos 2t + \tfrac{1}{3}B\sin 2t - \tfrac{1}{2}Pe^{2t} - \tfrac{1}{2}Qe^{-2t} + \tfrac{16}{75}e^t - \tfrac{1}{20}t.$$

NOTE. We might have anticipated that the number of arbitrary constants would be four, as the two second-order differential coefficients d^2x/dt^2, d^2y/dt^2 had to be 'integrated' in the process of solution. We managed to reduce the number from the initial eight to four by substitution in the first of the given equations; we might equally well have used the second, or, if necessary, both.

For Examples on the work of § 5, see Revision Examples XVII, nos. 50–65 (pp. 57–59).

6. The Euler linear equation. Closely allied to the linear differential equation with constant coefficients is the equation

$$a_0 x^n \frac{d^n y}{dx^n} + a_1 x^{n-1} \frac{d^{n-1} y}{dx^{n-1}} + \dots + a_{n-1} x \frac{dy}{dx} + a_n y = Q(x),$$

where $a_0, a_1, \dots, a_{n-1}, a_n$ are constants.

One or two methods of solution are available, but reduction to the 'constant coefficients' type will probably be found as useful as any. For this we make the substitution

$$x = e^t,$$

and then seek to express $x\dfrac{dy}{dx}, x^2\dfrac{d^2y}{dx^2}, \dots$ as functions of $\dfrac{dy}{dt}, \dfrac{d^2y}{dt^2}, \dots$. By direct differentiation, we have the relation

$$\frac{dy}{dt} = \frac{dy}{dx}\frac{dx}{dt} = \frac{dy}{dx}e^t$$

$$= x\frac{dy}{dx}.$$

Then
$$\frac{d^2y}{dt^2} = \frac{d}{dx}\left(x\frac{dy}{dx}\right)\frac{dx}{dt} = \left(x\frac{d^2y}{dx^2} + \frac{dy}{dx}\right)e^t$$

$$= x^2\frac{d^2y}{dx^2} + x\frac{dy}{dx}.$$

For equations of the second order, such as are most common at this stage, this analysis suffices, and we have the identities

$$x\frac{dy}{dx} \equiv \frac{dy}{dt},$$

$$x^2\frac{d^2y}{dx^2} \equiv \frac{d^2y}{dt^2} - \frac{dy}{dt},$$

reducing the given equation

$$a_0 x^2 \frac{d^2y}{dx^2} + a_1 x \frac{dy}{dx} + a_2 y = Q(x)$$

to the form
$$a_0 \frac{d^2y}{dt^2} + (a_1 - a_0)\frac{dy}{dt} + a_2 y = Q(e^t),$$

with which we have already learned to deal.

For equations of higher order, we may establish the more general formula:

If D is written for the operation d/dt, then

$$x^p \frac{d^py}{dx^p} \equiv D(D-1)(D-2)\dots(D-p+1)\,y.$$

This is easily verified by induction. For

$$\frac{d}{dt}\left(x^p \frac{d^py}{dx^p}\right) = \left(x^p \frac{d^{p+1}y}{dx^{p+1}} + p x^{p-1}\frac{d^py}{dx^p}\right)\frac{dx}{dt}$$
$$= x^{p+1}\frac{d^{p+1}y}{dx^{p+1}} + p x^p \frac{d^py}{dx^p},$$

since $dx/dt = e^t = x$. Hence

$$x^{p+1}\frac{d^{p+1}y}{dx^{p+1}} = D\left(x^p \frac{d^py}{dx^p}\right) - p\left(x^p \frac{d^py}{dx^p}\right)$$
$$= (D-p)\left(x^p \frac{d^py}{dx^p}\right).$$

If we assume the formula to be true for any definite value p, then

$$x^{p+1}\frac{d^{p+1}y}{dx^{p+1}} = (D-p)\,D(D-1)\dots(D-p+1)\,y$$
$$= D(D-1)\dots(D-p+1)(D-p)\,y,$$

and so it is true for $p+1$. But we have proved it for $p=1$; hence it holds for $p = 2, 3, 4, \dots$, and so generally.

ILLUSTRATION 18. *To solve the equation*

$$x^2\frac{d^2y}{dx^2} - 2x\frac{dy}{dx} + 2y = 2\log x.$$

Substitute
$$x = e^t.$$

Then
$$x\frac{dy}{dx} = \frac{dy}{dt},$$

$$x^2\frac{d^2y}{dx^2} = \frac{d^2y}{dt^2} - \frac{dy}{dt},$$

and the equation becomes

$$\frac{d^2y}{dt^2}-3\frac{dy}{dt}+2y=2t.$$

The solution of the equation in t is

$$y=Ae^t+Be^{2t}+t+\tfrac{3}{2},$$

where A, B are arbitrary constants. Hence the solution of the given equation is

$$y=Ax+Bx^2+\log x+\tfrac{3}{2}.$$

ALITER. The following alternative solution gives another example of the method of Variation of Parameters.

The equation

$$x^2\frac{d^2y}{dx^2}-2x\frac{dy}{dx}+2y=0$$

may be expected to have a solution of the form

$$y=ax^n,$$

for then $$y'=nax^{n-1},\quad y''=n(n-1)\,ax^{n-2},$$

and substitution gives the relation

$$n(n-1)\,ax^n-2nax^n+2ax^n=0,$$

or $$(n^2-3n+2)\,ax^n=0.$$

This is satisfied when $n=1$ or $n=2$.

Taking $n=1$, consider the solution $y=ax$. Make the substitution

$$y=zx$$

in the given equation. Then

$$y'=z'x+z,\quad y''=z''x+2z',$$

and the equation becomes

$$(z''x^3+2z'x^2)-2(z'x^2+zx)+2zx=2\log x,$$

or $$z''=2x^{-3}\log x.$$

Hence $$\begin{aligned} z' &= -x^{-2}\log x+\int x^{-2}.x^{-1}dx+\text{constant} \\ &= -x^{-2}\log x-\tfrac{1}{2}x^{-2}+A, \end{aligned}$$

so that $$\begin{aligned} z &= x^{-1}\log x-\int x^{-1}.x^{-1}dx+\tfrac{1}{2}x^{-1}+Ax+\text{constant} \\ &= x^{-1}\log x+\tfrac{3}{2}x^{-1}+Ax+B. \end{aligned}$$

Thus $$y=zx=\log x+\tfrac{3}{2}+Ax^2+Bx.$$

For Examples on the work of § 6, see Revision Examples XVII, nos. 66–72 (p. 59).

REVISION EXAMPLES XVII

Solve the differential equations:

1. $\dfrac{d^2y}{dx^2}+4\dfrac{dy}{dx}+13y=\sin 3x$, where $y=0$ when $x=0$ and $\frac{1}{2}\pi$.

2. $\dfrac{d^2y}{dx^2}+3\dfrac{dy}{dx}+2y=xe^{-x}$.

3. $\dfrac{d^2y}{dx^2}+5\dfrac{dy}{dx}+6y=e^{2x}$, where $y=0$, $\dfrac{dy}{dx}=1$ when $x=0$.

4. $\dfrac{d^2y}{dx^2}-n^2y=e^{nx}$.

5. $\dfrac{d^2y}{dx^2}+4\dfrac{dy}{dx}+5y=e^{-2x}$.

6. $\dfrac{d^2y}{dx^2}+\dfrac{dy}{dx}+y=\sin x$, where $y=0$, $\dfrac{dy}{dx}=1$ when $x=0$.

7. $\dfrac{d^2y}{dx^2}+y=2\cos^2 x$, where $y=0$, $\dfrac{dy}{dx}=0$ when $x=0$.

8. $\dfrac{d^2y}{dx^2}+4\dfrac{dy}{dx}+5y=\sin 2x$.

9. $\dfrac{d^2y}{dx^2}+4\dfrac{dy}{dx}+4y=2\cos^2 x$.

10. $\dfrac{d^2y}{dx^2}+5\dfrac{dy}{dx}+6y=e^x(x+1)$, where $y=1$, $\dfrac{dy}{dx}=0$ when $x=0$.

11. $\dfrac{d^2y}{dx^2}-6\dfrac{dy}{dx}+10y=20-e^{2x}$.

12. $\dfrac{d^2y}{dx^2}+6\dfrac{dy}{dx}+9y=27x$, where $y=\dfrac{dy}{dx}=0$ when $x=0$.

13. $\dfrac{d^3y}{dx^3}-2\dfrac{d^2y}{dx^2}+\dfrac{dy}{dx}-2y=12\sin 2x-4x$, where $y=2$, $\dfrac{dy}{dx}=5$, $\dfrac{d^2y}{dx^2}=-4$ when $x=0$.

14. $4\dfrac{d^2y}{dx^2}-8\dfrac{dy}{dx}-5y=9xe^x$.

15. $\dfrac{d^2y}{dx^2}+2\dfrac{dy}{dx}+2y=5\sin x$, where $y=0$, $\dfrac{dy}{dx}=0$ when $x=0$.

16. $\dfrac{d^2y}{dx^2}-y=x\sinh x$.

17. $\dfrac{d^2y}{dx^2}-4\dfrac{dy}{dx}+4y=e^{2x}\sin x$.

18. $\frac{d^2y}{dx^2}+6\frac{dy}{dx}+9y=\sin x$, where $y=1$, $\frac{dy}{dx}=0$ when $x=0$.

19. $\frac{d^2y}{dx^2}+3\frac{dy}{dx}+2y=e^{-x}$, where $y=0$, $\frac{dy}{dx}=0$ when $x=0$.

20. $\frac{d^2x}{dt^2}=-n^2(x-\frac{1}{2}ft^2)$, where $x=a$, $\frac{dx}{dt}=0$ when $t=0$.

21. $\frac{d^2x}{dt^2}+2k\frac{dx}{dt}+n^2x=0$ $(n>k)$, where $x=a$, $\frac{dx}{dt}=0$ when $t=0$.

22. $\frac{d^2y}{dx^2}-\frac{dy}{dx}=e^{-x}$, where $y=0$ when $x=0$, and where y is positive for all other values of x (positive or negative).

23. $\frac{d^2y}{dx^2}+2\frac{dy}{dx}+y=e^{-x}\sin^2 x$.

24. $\frac{d^2y}{dx^2}-4\frac{dy}{dx}+5y=x^2+\cos 2x$.

25. $\frac{d^2y}{dx^2}+2\frac{dy}{dx}+y=\sin x$, where $y=0$, $\frac{dy}{dx}=0$ when $x=0$.

26. $\frac{d^2y}{dx^2}+y=\sin x$, where $y=0$, $\frac{dy}{dx}=0$ when $x=0$.

27. $\frac{d^2y}{dx^2}-3\frac{dy}{dx}+2y=e^x$, where $y=3$, $\frac{dy}{dx}=3$ when $x=0$.

28. $\frac{d^2y}{dx^2}-y=1$, where $y=0$ when $x=0$ and where y tends to a finite limit as $x\to-\infty$.

29. $\frac{d^3y}{dx^3}-\frac{d^2y}{dx^2}+\frac{dy}{dx}-y=0$.

30. $\frac{d^2x}{dt^2}+2\frac{dx}{dt}+17x=5\sin 4t$.

31. $2\frac{d^2y}{dx^2}-3\frac{dy}{dx}+y=-e^x$, given that $y=1$ when $x=0$, and that $y=0$ when $x=1$.

32. $\frac{d^2y}{dx^2}+6\frac{dy}{dx}+9y=x$.

33. $\frac{d^2y}{dx^2}+4\frac{dy}{dx}+3y=\sin x$.

34. $\frac{d^2y}{dx^2}+4\frac{dy}{dx}+8y=1+\sin x\cos x$.

35. $\dfrac{d^2y}{dx^2}+2\dfrac{dy}{dx}+3y=\cos x.$

36. $4\dfrac{d^2y}{dx^2}-12\dfrac{dy}{dx}+9y=xe^{\frac{3}{2}x}.$

37. $\dfrac{d^2y}{dx^2}-5\dfrac{dy}{dx}+6y=5(\sin x-\cos x).$

38. $\dfrac{d^2y}{dx^2}+3\dfrac{dy}{dx}+2y=2e^{-2x}+e^{2x}.$

39. $\dfrac{d^2y}{dx^2}+3\dfrac{dy}{dx}=x+x^2.$

40. $\dfrac{d^2y}{dx^2}+4\dfrac{dy}{dx}+4y=e^{2x}.$

41. $\dfrac{d^2y}{dx^2}+6\dfrac{dy}{dx}+25y=104e^{3x}$, given that $y=2$ when $x=0$, and that $y=0$ when $x=\frac{1}{8}\pi$.

42. $\dfrac{d^2y}{dx^2}+y=\cos x.$

43. $\dfrac{d^2y}{dx^2}+5\dfrac{dy}{dx}+6y=e^{-2x}\sin 2x.$

44. $\dfrac{d^4y}{dx^4}+2\dfrac{d^2y}{dx^2}+y=x\sin x+\cos^2 x.$

45. $\dfrac{d^2y}{dx^2}+2\dfrac{dy}{dx}+y=x+e^{-x}.$

46. $\dfrac{d^3y}{dx^3}+a^2\dfrac{dy}{dx}=4x\cos ax.$

47. $\dfrac{d^3y}{dx^3}-2\dfrac{d^2y}{dx^2}-19\dfrac{dy}{dx}+20y=xe^x+2e^{-4x}\sin x.$

48. $\dfrac{d^2y}{dx^2}+4y=8x\cos 2x+2\sin 2x.$

49. $\dfrac{d^3y}{dx^3}-2\dfrac{d^2y}{dx^2}-\dfrac{dy}{dx}+2y=12(x+\cosh x).$

Solve the simultaneous differential equations:

50. $\dfrac{dx}{dt}+4\dfrac{dy}{dt}+3x=3t+3,$

$$2\frac{dx}{dt}-\frac{dy}{dt}+3y=6t-3,$$

where $x=0$, $y=0$ when $t=0$. Show that $x=2e^{-1}-1$ when $t=1$ and find the corresponding value of y.

51. $\dfrac{dx}{dt}+\dfrac{dy}{dt}-3x=e^{-2t},\qquad \dfrac{dx}{dt}-\dfrac{dy}{dt}-2y=e^{2t},$

where $x=0$, $y=0$ when $t=0$.

52. $\dfrac{dx}{dt}+y-z=\cos t,\quad \dfrac{dy}{dt}+z-x=t,$

$\dfrac{dz}{dt}+x-y=0,$

where $x=1$, $y=\frac{1}{3}$, $z=1$ when $t=0$.

53. $\dfrac{dx}{dt}+x+2\dfrac{dy}{dt}+y=6te^{-2t},\quad \dfrac{dx}{dt}+\dfrac{dy}{dt}-4y=0,$

where $x=1$, $\dfrac{dx}{dt}=-8$ when $t=0$.

54. $\dfrac{dx}{dt}+x+\dfrac{dy}{dt}+y=t,\quad \dfrac{dx}{dt}+3x+2\dfrac{dy}{dt}+y=3t.$

55. $2\dfrac{dx}{dt}+3\dfrac{dy}{dt}-3x-2y=e^{t},\quad 4\dfrac{dx}{dt}+3\dfrac{dy}{dt}-4x-3y=e^{2t},$

where $x=1$, $y=0$ when $t=0$.

56. $\dfrac{dx}{dt}-ay=\sin at\quad (a\neq 0),\quad \dfrac{dy}{dt}+ax=\cos at,$

where $x=x_0$, $y=y_0$ when $t=0$.

57. $5\dfrac{d^2x}{dt^2}+\dfrac{dy}{dt}+2x=4\cos t,\quad 3\dfrac{dx}{dt}+y=8t\cos t,$

where $x=1$, $\dfrac{dx}{dt}=0$ when $t=0$.

58. $\dfrac{d^2x}{dt^2}+3\dfrac{dx}{dt}-2x+\dfrac{dy}{dt}-3y=2e^{-t},\quad 2\dfrac{dx}{dt}-x+\dfrac{dy}{dt}-2y=0,$

where $x=0$, $\dfrac{dx}{dt}=0$, $y=4$ when $t=0$.

59. $2\dfrac{d^2y}{dx^2}-\dfrac{dz}{dx}-4y=2x,\quad 2\dfrac{dy}{dx}+4\dfrac{dz}{dx}-3z=0,$

where $y=1$, $\dfrac{dy}{dx}=\frac{1}{2}$, $z=-\frac{7}{3}$ when $x=0$.

60. $\dfrac{d^2y}{dt^2}+2\dfrac{d^2x}{dt^2}=2y-3x-1,\quad \dfrac{d^2x}{dt^2}-2\dfrac{d^2y}{dt^2}=x+16y-3.$

61. $\dfrac{d^2x}{dt^2} - y = e^{2t}, \qquad 2\dfrac{d^2x}{dt^2} - \dfrac{d^2y}{dt^2} - x = 4e^{2t},$

where $x = \dfrac{dx}{dt} = \dfrac{d^2x}{dt^2} = 0,\ \dfrac{dy}{dt} = -6$ when $t = 0$.

62. $\dfrac{d^2x}{dt^2} + \dfrac{d^2y}{dt^2} - 3x + y = \sinh t, \qquad \dfrac{d^2x}{dt^2} - y = \sin(t\sqrt{3}).$

63. $\dfrac{d^2y}{dt^2} - 2\dfrac{dy}{dt} - 2y + 3x = 0, \qquad \dfrac{dx}{dt} + \dfrac{dy}{dt} - 6x + 4y = -7,$

where $y = -12,\ x = 7,\ \dfrac{dy}{dx} = 0$ when $t = 0$.

64. $\dfrac{d^2x}{dt^2} - \dfrac{dy}{dt} + x = 0, \qquad \dfrac{d^2y}{dt^2} + \dfrac{dx}{dt} + y = e^t.$

65. $\dfrac{d^2x}{dt^2} + 2\dfrac{dy}{dt} - x + \sin t = 0, \qquad \dfrac{d^2y}{dt^2} - 2\dfrac{dx}{dt} - y - \cos t = 0.$

Solve the differential equations:

66. $x^2\dfrac{d^2y}{dx^2} + x\dfrac{dy}{dx} + y = x.$

67. $x^2\dfrac{d^2y}{dx^2} - 2x\dfrac{dy}{dx} + 2y = x^3.$

68. $x^2\dfrac{d^2y}{dx^2} - 4x\dfrac{dy}{dx} + 6y = x.$

69. $x^2\dfrac{d^2y}{dx^2} + 8x\dfrac{dy}{dx} + 12y = x^{-3}\log x + 252.$

70. $x^2\dfrac{d^2y}{dx^2} - 3x\dfrac{dy}{dx} + y = x^4.$

71. $x^2\dfrac{d^2y}{dx^2} - 2x\dfrac{dy}{dx} + 2y = x^3\cos x.$

72. $x^3\dfrac{d^3y}{dx^3} + 2x^2\dfrac{d^2y}{dx^2} + 3x\dfrac{dy}{dx} - 3y = 7x^2.$

Solve the simultaneous differential equations:

73. $t\dfrac{d^2x}{dt^2} + \dfrac{dx}{dt} + \dfrac{dy}{dt} = 1, \qquad t\dfrac{dy}{dt} - 4t\dfrac{dx}{dt} + 7x + 2y = t.$

74. $t^2\dfrac{d^2x}{dt^2} - t\dfrac{dy}{dt} + y = t, \qquad t^2\dfrac{d^2y}{dt^2} + \dfrac{dx}{dt} - x = t^2.$

75. Show that, provided $p^2 > 4q$, the differential equation

$$x^2\frac{d^2y}{dx^2}+(p+1)x\frac{dy}{dx}+qy=0$$

has, for $x>0$, two distinct solutions of the form $y=x^c$, where c is real. Hence write down the general solution.

Find the solution of the equation

$$\frac{d^2y}{dx^2}-\frac{6}{x^2}y=0,$$

where $y=1$ when $x=1$, and where $y\to 0$ as $x\to\infty$.

76. Transform the equation

$$x^2\frac{d^2y}{dx^2}-(1-3x)\frac{dy}{dx}+y=0$$

by putting $x=1/t$, $y=ute^{-t}$.

Hence, or otherwise, find a solution which remains finite when x tends to infinity

77. If
$$f(x)=\phi(x)\cos x+\psi(x)\sin x,$$
$$g(x)=\psi(x)\cos x-\phi(x)\sin x,$$

where $\phi(x)$, $\psi(x)$ are differentiable functions, express $\phi'(x)$, $\psi'(x)$ in terms of $f(x)$, $f'(x)$, $g(x)$, $g'(x)$.

Deduce that *every* function satisfying the relation

$$f''(x)=-f(x)$$

throughout a range $a\leqslant x\leqslant b$ is (in that range) of the form

$$L\cos x+M\sin x,$$

where L, M are constants.

CHAPTER XXII

THE LINEAR DIFFERENTIAL EQUATION WITH CONSTANT COEFFICIENTS; ALTERNATIVE METHOD

1. The linear equation of the first order. We begin by retracing the argument to consider the equation

$$\frac{dy}{dx} - py = Q(x),$$

where p is constant and $Q(x)$ is a function of x only. Multiplying by e^{-px} as an integrating factor, we obtain the equation in the form

$$e^{-px}\frac{dy}{dx} - pe^{-px}\,y = Q(x)\,e^{-px},$$

or
$$\frac{d}{dx}(ye^{-px}) = Q(x)\,e^{-px}.$$

If we integrate from an arbitrary lower limit a, and denote the current variable of integration by the letter t, we obtain a solution of the equation in the form (compare Volume I, p. 85)

$$\Big[y(t)\,e^{-pt}\Big]_a^x = \int_a^x Q(t)\,e^{-pt}\,dt,$$

or
$$y(x)\,e^{-px} - y(a)\,e^{-pa} = \int_a^x Q(t)\,e^{-pt}\,dt.$$

Hence
$$y(x) = y(a)\,e^{p(x-a)} + \int_a^x Q(t)\,e^{p(x-t)}\,dt,$$

where, it will be remembered, x is constant for the purposes of the integration on the right-hand side.

Since the constant a is arbitrary, this solution is exhibited in standard form with

$$y(a)\,e^{p(x-a)}$$

as complementary function and

$$\int_a^x Q(t)\,e^{p(x-t)}\,dt$$

as particular integral. We have found on the way the special solution which arises if y has a given value $y(a)$ when $x = a$.

In particular, *the solution which satisfies the condition*

$$y=0 \quad when \quad x=a$$

is
$$y=\int_a^x Q(t)\, e^{p(x-t)}\, dt.$$

The usefulness of this form of solution depends on our ability to produce a 'calculus' for the function $\int_a^x Q(t)\, e^{p(x-t)}\, dt$ as convenient to manipulate as the alternatives given in the preceding chapter. This will appear later, but we first extend the result to equations of higher order.

ILLUSTRATION 1. *To solve the equation*

$$\frac{dy}{dx}-2y=e^{2x}\sin x,$$

given that $y=3$ when $x=\frac{1}{2}\pi$.

The equation is
$$\frac{d}{dx}(ye^{-2x})=\sin x.$$

Integrate from $\frac{1}{2}\pi$ to x, and denote the current variable of integration by the letter t. Then

$$\Big[y(t)\, e^{-2t}\Big]_{\frac{1}{2}\pi}^{x}=\Big[-\cos t\Big]_{\frac{1}{2}\pi}^{x},$$

or
$$y(x)\, e^{-2x}-3e^{-\pi}=-\cos x,$$

so that
$$y=3e^{2x-\pi}-e^{2x}\cos x.$$

2. The linear equation of the second order; auxiliary roots unequal. If p, q are the roots of the auxiliary equation, then the given equation can be expressed in the form

$$\frac{d^2y}{dx^2}-(p+q)\frac{dy}{dx}+pqy=Q(x).$$

We assume that p, q are unequal. It is also supposed that the values of y, y' are known when x has the value a; say

$$y=y_1, \quad y'=y_1' \quad \text{when} \quad x=a.$$

The given equation is equivalent to

$$(D-p)(D-q)\,y=Q(x),$$

and, if we write
$$(D-q)\,y\equiv u,$$

then the function $u(x)$ satisfies the linear equation

$$\frac{du}{dx} - pu = Q(x).$$

Hence, by the preceding paragraph,

$$u(x) = u(a)\, e^{p(x-a)} + \int_a^x Q(t)\, e^{p(x-t)}\, dt$$

on taking as the lower limit of integration the value a, for which y, y' are known. Since

$$u(a) \equiv y'(a) - qy(a),$$

this solution is

$$u(x) = (y_1' - qy_1)\, e^{p(x-a)} + \int_a^x Q(t)\, e^{p(x-t)}\, dt.$$

Thus, by definition of $u(x)$,

$$y' - qy = (y_1' - qy_1)\, e^{p(x-a)} + \int_a^x Q(t)\, e^{p(x-t)}\, dt.$$

The whole argument may be repeated with the roles of p, q interchanged, and this leads to the relation

$$y' - py = (y_1' - py_1)\, e^{q(x-a)} + \int_a^x Q(t)\, e^{q(x-t)}\, dt.$$

The value of y is obtained at once by subtracting these two equations and then dividing by $p-q$, which is not zero.

In particular, *the solution which satisfies the conditions*

$$y = 0, \quad y' = 0 \quad \textit{when} \quad x = a$$

is

$$y = \frac{1}{p-q}\int_a^x Q(t)\, e^{p(x-t)}\, dt + \frac{1}{q-p}\int_a^x Q(t)\, e^{q(x-t)}\, dt$$

$$= \frac{1}{p-q}\int_a^x \{e^{p(x-t)} - e^{q(x-t)}\}\, Q(t)\, dt.$$

COROLLARY. Since we know in any case that the complementary function is

$$Ae^{px} + Be^{qx},$$

the general solution of the given equation may be expressed in the form (with lower limit zero for the particular integral)

$$y = Ae^{px} + Be^{qx} + \frac{1}{p-q}\int_0^x \{e^{p(x-t)} - e^{q(x-t)}\}\, Q(t)\, dt.$$

ILLUSTRATION 2. *To solve the differential equation*

$$\frac{d^2y}{dx^2}-5\frac{dy}{dx}+6y=xe^x,$$

given that $y=1$, $y'=2$ *when* $x=0$.

We follow the details of the solution given in the text. A more condensed treatment will be given in the next illustration.

The auxiliary equation is

$$p^2-5p+6=0,$$

with roots 2 and 3, and the given equation is

$$(D-2)(D-3)y=xe^x.$$

Write $$(D-3)y\equiv u;$$
then the given equation is

$$(D-2)u=xe^x,$$

and the solution is
$$\begin{aligned}u&=u_0e^{2x}+\int_0^x te^t e^{2(x-t)}dt\\&=u_0e^{2x}+e^{2x}\int_0^x te^{-t}dt\\&=u_0e^{2x}+e^{2x}\Big[-(1+t)e^{-t}\Big]_0^x.\end{aligned}$$

Hence
$$\begin{aligned}y'-3y&=(y_0'-3y_0)e^{2x}+e^{2x}\{-(1+x)e^{-x}+1\}\\&=(2-3)e^{2x}-(1+x)e^x+e^{2x}\end{aligned}$$

in virtue of the given conditions. It follows that

$$y'-3y=-(1+x)e^x.$$

(This equation may now be integrated by the same method; but, to illustrate the text, we make a fresh start.)

Next write $$(D-2)y=v;$$
then the given equation is

$$(D-3)v=xe^x,$$

and the solution is

$$\begin{aligned}v&=v_0e^{3x}+\int_0^x te^t e^{3(x-t)}dt\\&=v_0e^{3x}+e^{3x}\int_0^x te^{-2t}dt\\&=v_0e^{3x}+e^{3x}\Big[-\tfrac{1}{4}(1+2t)e^{-2t}\Big]_0^x.\end{aligned}$$

Hence $$y' - 2y = (y_0' - 2y_0)\,e^{3x} + e^{3x}\{-\tfrac{1}{4}(1+2x)\,e^{-2x} + \tfrac{1}{4}\}$$
$$= (2-2)\,e^{3x} - \tfrac{1}{4}(1+2x)\,e^{x} + \tfrac{1}{4}e^{3x}$$
$$= \tfrac{1}{4}e^{3x} - \tfrac{1}{4}(1+2x)\,e^{x}.$$

Subtracting the equation for $y' - 3y$ from the equation for $y' - 2y$, we obtain the required solution

$$y = \tfrac{1}{4}e^{3x} - \tfrac{1}{4}(1+2x)\,e^{x} + (1+x)\,e^{x}$$
$$= \tfrac{1}{4}e^{3x} + \tfrac{1}{4}(3+2x)\,e^{x}.$$

ILLUSTRATION 3. *To solve the differential equation*

$$\frac{d^2y}{dx^2} - 3\frac{dy}{dx} + 2y = 3x^2e^{2x}.$$

This equation was solved on p. 45, and the two methods should be compared.

Taking $p=2$, $q=1$ in the formula

$$y = \frac{1}{p-q}\int_0^x \{e^{p(x-t)} - e^{q(x-t)}\}\,Q(t)\,dt,$$

we obtain the particular integral

$$y = \int_0^x \{e^{2(x-t)} - e^{(x-t)}\}\,3t^2e^{2t}\,dt$$
$$= e^{2x}\int_0^x 3t^2\,dt - e^{x}\int_0^x 3t^2e^{t}\,dt$$
$$= e^{2x}\Big[\,t^3\,\Big]_0^x - e^{x}\Big[\,(3t^2 - 6t + 6)\,e^{t}\,\Big]_0^x$$
$$= x^3e^{2x} - (3x^2 - 6x + 6)\,e^{2x} + 6e^{x}$$
$$= (x^3 - 3x^2 + 6x - 6)\,e^{2x} + 6e^{x}.$$

The general solution, on absorbing the terms $-6e^{2x}$ and $+6e^{x}$ into the arbitrary constants of the complementary function, is

$$y = Ae^{2x} + Be^{x} + (x^3 - 3x^2 + 6x)\,e^{2x}.$$

ILLUSTRATION 4. *To solve the differential equation*

$$\frac{d^2y}{dx^2} - 3\frac{dy}{dx} + 2y = \begin{cases} 3 & \text{when} \quad x < 1, \\ 7 & \text{when} \quad x > 1, \end{cases}$$

under the conditions

$$y = y' = 0 \quad \textit{when} \quad x = -1.$$

This is the usual form of wording for such problems. It is implicit that y and y' are to be regarded as continuous functions, but note that y'' *does not then exist at the actual 'break'* since its values are different for approach along $x<1$ and $x>1$.

Taking $p=2$, $q=1$, we have the solution

$$y=\int_{-1}^{x}\{e^{2(x-t)}-e^{(x-t)}\}\,Q(t)\,dt,$$

where

$$Q(t)\equiv\begin{cases}3 & \text{when} \quad x<1,\\ 7 & \text{when} \quad x>1.\end{cases}$$

When $x<1$, the solution is

$$y_{x<1}=\int_{-1}^{x}3\{e^{2(x-t)}-e^{(x-t)}\}\,dt,$$

since $Q(t)\equiv 3$ in that range. Hence

$$\begin{aligned}y_{x<1}&=-\tfrac{3}{2}e^{2x}\Big[e^{-2t}\Big]_{-1}^{x}+3e^{x}\Big[e^{-}\Big]_{-1}^{x}\\ &=-\tfrac{3}{2}e^{2x}(e^{-2x}-e^{2})+3e^{x}(e^{-x}-e)\\ &=\tfrac{3}{2}e^{2(x+1)}-3e^{(x+1)}+\tfrac{3}{2}.\end{aligned}$$

When $x>1$, we must split up the interval of integration; thus

$$\begin{aligned}y_{x>1}&=\int_{-1}^{1}3\{e^{2(x-t)}-e^{(x-t)}\}\,dt+7\int_{1}^{x}\{e^{2(x-t)}-e^{(x-t)}\}\,dt\\ &=-\tfrac{3}{2}e^{2x}\Big[e^{-2t}\Big]_{-1}^{1}+3e^{x}\Big[e^{-t}\Big]_{-1}^{1}-\tfrac{7}{2}e^{2x}\Big[e^{-2t}\Big]_{1}^{x}+7e^{x}\Big[e^{-t}\Big]_{1}^{x}\\ &=-\tfrac{3}{2}e^{2x}(e^{-2}-e^{2})+3e^{x}(e^{-1}-e)-\tfrac{7}{2}e^{2x}(e^{-2x}-e^{-2})+7e^{x}(e^{-x}-e^{-1})\\ &=e^{2x}(2e^{-2}+\tfrac{3}{2}e^{2})-e^{x}(4e^{-1}+3e)+\tfrac{7}{2}.\end{aligned}$$

NOTE. The method which we are elaborating is particularly useful when $Q(x)$ has the form exhibited in this illustration. It may be helpful to give a solution by the earlier method also.

When $x<1$, the solution assumes the form

$$y=Ae^{2x}+Be^{x}+\tfrac{3}{2},$$

where, from the condition '$y=y'=0$ when $x=-1$',

$$0=Ae^{-2}+Be^{-1}+\tfrac{3}{2},$$

$$0=2Ae^{-2}+Be^{-1}.$$

Hence $$Ae^{-2} = \tfrac{3}{2}, \quad Be^{-1} = -3,$$

so that
$$\begin{aligned} y &= \tfrac{3}{2}e^2\, e^{2x} - 3e\, e^x + \tfrac{3}{2} \\ &= \tfrac{3}{2}e^{2(x+1)} - 3e^{(x+1)} + \tfrac{3}{2}. \end{aligned}$$

When $x > 1$, the solution is

$$y = Pe^{2x} + Qe^x + \tfrac{7}{2},$$

where the arbitrary constants may be expected to be different from those for the case $x < 1$.

In order to find P and Q, observe that, since (p. 66) the functions y and y' are to be taken as continuous,

$$\begin{aligned} Pe^2 + Qe + \tfrac{7}{2} &= \tfrac{3}{2}e^4 - 3e^2 + \tfrac{3}{2}, \\ 2Pe^2 + Qe \quad &= 3e^4 - 3e^2. \end{aligned}$$

Subtracting, we have
$$Pe^2 - \tfrac{7}{2} = \tfrac{3}{2}e^4 - \tfrac{3}{2},$$

or $$P = \tfrac{3}{2}e^2 + 2e^{-2}.$$

Hence $$Q = -(3e + 4e^{-1}).$$

The solution for $x > 1$ is therefore

$$y = (\tfrac{3}{2}e^2 + 2e^{-2})\, e^{2x} - (3e + 4e^{-1})\, e^x + \tfrac{7}{2}.$$

3. The linear equation of the second order; auxiliary roots complex. If the given equation is

$$a_0 \frac{d^2y}{dx^2} + a_1 \frac{dy}{dx} + a_2 y = Q(x),$$

the auxiliary equation is

$$a_0 p^2 + a_1 p + a_2 = 0,$$

and we suppose that the auxiliary roots are the complex numbers $\alpha \pm i\beta$, where
$$\begin{aligned} 2\alpha &= -a_1/a_0, \\ \alpha^2 + \beta^2 &= \quad a_2/a_0. \end{aligned}$$

A particular solution, namely, that for which

$$y = 0, \quad y' = 0 \quad \text{when} \quad x = a,$$

is, by the preceding paragraph,

$$y=\frac{1}{2i\beta}\int_a^x \{e^{(\alpha+i\beta)(x-t)}-e^{(\alpha-i\beta)(x-t)}\}\,Q(t)\,dt$$

$$=\frac{1}{\beta}\int_a^x e^{\alpha(x-t)}\sin\beta(x-t)\,Q(t)\,dt.$$

COROLLARY. *The solution of the equation*

$$\frac{d^2y}{dx^2}+n^2y=Q(x)$$

which satisfies the conditions

$$y=0,\quad y'=0\quad \textit{when}\quad x=a,$$

is

$$y=\frac{1}{n}\int_a^x \sin n(x-t)\,Q(t)\,dt.$$

4. The linear equation of the second order; auxiliary roots equal. When the roots of the auxiliary equation are equal, the given equation can be expressed in the form

$$\frac{d^2y}{dx^2}-2p\frac{dy}{dx}+p^2y=Q(x).$$

We suppose, as before, that a particular solution is sought for which

$$y=0,\quad y'=0\quad \text{when}\quad x=a.$$

Writing

$$(D-p)\,y\equiv u,$$

we have the equation

$$(D-p)\,u=Q(x),$$

so that

$$u=\int_a^x Q(t)\,e^{p(x-t)}\,dt.$$

The equation for y is therefore

$$(D-p)\,y=R(x),$$

where

$$R(x)=\int_a^x Q(t)\,e^{p(x-t)}\,dt,$$

and so

$$y=\int_a^x R(t)\,e^{p(x-t)}\,dt$$

$$=e^{px}\int_a^x R(t)\,e^{-pt}\,dt.$$

In order to obtain a more convenient form, not involving repeated integration, we integrate by parts the expression

$$\int_a^x R(t)\, e^{-pt}\, dt,$$

giving
$$\Big[tR(t)\, e^{-pt}\Big]_a^x - \int_a^x t\{R'(t)\, e^{-pt} - pR(t)\, e^{-pt}\}\, dt.$$

But
$$R(t) = \int_a^t Q(\theta)\, e^{p(t-\theta)}\, d\theta$$
$$= e^{pt}\int_a^t Q(\theta)\, e^{-p\theta}\, d\theta,$$

so that

(i) $R(a) = 0$,

(ii) $R'(t) = pe^{pt}\int_a^t Q(\theta)\, e^{-p\theta}\, d\theta + e^{pt}\, Q(t)\, e^{-pt}$

$\qquad = pR(t) + Q(t).$

Hence
$$\int_a^x R(t)\, e^{-pt}\, dt = xR(x)\, e^{-px} - \int_a^x t\, Q(t)\, e^{-pt}\, dt$$
$$= xe^{-px}\int_a^x Q(t)\, e^{p(x-t)}\, dt - \int_a^x t\, Q(t)\, e^{-pt}\, dt$$
$$= \int_a^x (x-t)\, Q(t)\, e^{-pt}\, dt.$$

We therefore have the particular integral in the simple form

$$y = \int_a^x (x-t)\, Q(t)\, e^{p(x-t)}\, dt.$$

ILLUSTRATION 5. *To solve the differential equation*

$$\frac{d^2y}{dx^2} - 4\frac{dy}{dx} + 4y = 12x^2e^{2x}.$$

The particular solution such that

$$y = 0, \quad y' = 0 \quad \text{when} \quad x = 0$$

is
$$y = \int_0^x 12(x-t)\, t^2e^{2t}\, e^{2(x-t)}\, dt$$
$$= e^{2x}\int_0^x 12(xt^2 - t^3)\, dt$$
$$= e^{2x}\Big[4xt^3 - 3t^4\Big]_0^x$$
$$= x^4e^{2x}.$$

Hence the general solution is

$$y = (A + Bx)\, e^{2x} + x^4e^{2x}.$$

5. The evaluation of $\int_a^x Q(t)\, e^{p(x-t)}\, dt$. If the method indicated in this chapter is to be helpful, means must be found for the ready evaluation of the expression

$$\int_a^x Q(t)\, e^{p(x-t)}\, dt.$$

The integral is capable of any degree of complexity according to the form of $Q(x)$ and, in the last resort, its evaluation must follow the customary processes for definite integration. In normal cases, however, when $Q(t)$ is of the type studied in the preceding chapter, the work is relatively easy.

If $Q(t)$ is an exponential, or reducible to exponential form (as for sines and cosines), the integration is immediate. There is also a simple rule which is convenient when $Q(t)$ is a polynomial, possibly residual to an exponential function; say

$$Q(t) \equiv e^{kt} f(t),$$

where $f(t)$ is a polynomial of degree m. We have then to evaluate

$$\int_a^x e^{kt} f(t)\, e^{p(x-t)}\, dt,$$

or

$$e^{px} \int_a^x f(t)\, e^{(k-p)t}\, dt.$$

Consider, then, the expression

$$\int_a^x f(t)\, e^{rt}\, dt \quad (r = k - p).$$

By continued integration by parts, we have

$$\begin{aligned}\int_a^x f e^{rt}\, dt &= \frac{1}{r}\Big[f e^{rt}\Big]_a^x - \frac{1}{r}\int_a^x f' e^{rt}\, dt \\ &= \frac{1}{r}\Big[f e^{rt}\Big]_a^x - \frac{1}{r^2}\Big[f' e^{rt}\Big]_a^x + \frac{1}{r^2}\int_a^x f'' e^{rt}\, dt \\ &= \frac{1}{r}\Big[f e^{rt}\Big]_a^x - \frac{1}{r^2}\Big[f' e^{rt}\Big]_a^x + \frac{1}{r^3}\Big[f'' e^{rt}\Big]_a^x - \frac{1}{r^3}\int_a^x f''' e^{rt}\, dt,\end{aligned}$$

and so on. The process ultimately stops, since $f^{(m+1)}$ vanishes identically, so that

$$\int_a^x fe^{rt}\,dt = \frac{1}{r}e^{rx}\left\{f(x)-\frac{f'(x)}{r}+\frac{f''(x)}{r^2}-\dots+(-)^m\frac{f^{(m)}(x)}{r^m}\right\}$$
$$-\frac{1}{r}e^{ra}\left\{f(a)-\frac{f'(a)}{r}+\frac{f''(a)}{r^2}-\dots+(-)^m\frac{f^{(m)}(a)}{r^m}\right\}.$$

Hence *the expression* $\int_a^x Q(t)\,e^{p(x-t)}\,dt$ *may be found from the formula*

$$\int_a^x Q(t)\,e^{p(x-t)}\,dt \equiv \int_a^x e^{kt}f(t)\,e^{p(x-t)}\,dt$$
$$=\frac{1}{r}e^{kx}\left\{f(x)-\frac{f'(x)}{r}+\frac{f''(x)}{r^2}-\dots+(-)^m\frac{f^{(m)}(x)}{r^m}\right\}$$
$$-\frac{1}{r}e^{px+(k-p)a}\left\{f(a)-\frac{f'(a)}{r}+\frac{f''(a)}{r^2}-\dots+(-)^m\frac{f^{(m)}(a)}{r^m}\right\},$$

where $r=k-p$.

ILLUSTRATION 6. *To solve the equation*

$$\frac{d^2y}{dx^2}-5\frac{dy}{dx}+6y=x^4e^{2x},$$

given that $\quad y=0,\quad y'=0\quad$ *when* $\quad x=0.$

The solution is $\quad \displaystyle\int_0^x \{e^{3(x-t)}-e^{2(x-t)}\}\,t^4e^{2t}\,dt.$

For $\displaystyle\int_0^x e^{3(x-t)}t^4e^{2t}\,dt$, we have $p=3$, $k=2$, so that its value is

$$-1\,.\,e^{2x}\left\{x^4-\frac{4x^3}{(-1)}+\frac{12x^2}{(-1)^2}-\frac{24x}{(-1)^3}+\frac{24}{(-1)^4}\right\}$$
$$-(-1)\,e^{3x}\left\{0-\frac{4.0^3}{(-1)}+\frac{12.0^2}{(-1)^2}-\frac{24.0}{(-1)^3}+\frac{24}{(-1)^4}\right\},$$

or
$$-e^{2x}(x^4+4x^3+12x^2+24x+24)+24e^{3x}.$$

The value of $\displaystyle\int_0^x e^{2(x-t)}t^4e^{2t}\,dt$ may be obtained directly, namely,

$$e^{2x}\int_0^x t^4\,dt=\tfrac{1}{5}x^5e^{2x}.$$

In all, then,

$$y=-e^{2x}\left(\tfrac{1}{5}x^5+x^4+4x^3+12x^2+24x+24\right)+24e^{3x}.$$

NOTE. We have confined ourselves to equations of the second order, but the work can be extended if desired. The algebraic manipulations have to be varied at some points in the extensions.

EXAMPLES I

Any of the second-order equations given in the preceding chapter may be solved by these methods. The examples which follow are directed more specifically towards this form of solution.

1. Prove that the solution of the differential equation

$$\frac{dy}{dx}+ay=f(x) \quad (a \text{ constant}),$$

which vanishes at $x=0$, is

$$y=\int_0^x e^{-a(x-t)}f(t)\,dt.$$

Illustrate this result by considering the case in which $a=2$ and

$$f(x)=\begin{cases}0 & (x<1),\\ x-1 & (1\leqslant x\leqslant 2),\\ 1 & (2<x).\end{cases}$$

2. Verify by substitution, or otherwise, that

$$y=-\frac{1}{a-b}\int_0^x \{e^{-a(x-t)}-e^{-b(x-t)}\}f(t)\,dt$$

is the solution of

$$\frac{d^2y}{dx^2}+(a+b)\frac{dy}{dx}+aby=f(x) \quad (a,\ b \text{ unequal constants}),$$

for which $y=0$, $\dfrac{dy}{dx}=0$ when $x=0$.

Deduce that the solution of

$$\frac{d^2y}{dx^2}+11\frac{dy}{dx}+30y=f(x),$$

where
$$f(x)=\begin{cases}1 & (0<x<1),\\ 0 & (\text{otherwise}),\end{cases}$$

that satisfies $y=0$, $\dfrac{dy}{dx}=0$ when $x=0$ is given by

$$y=\frac{e^5-1}{5}e^{-5x}-\frac{e^6-1}{6}e^{-6x}$$

for values of x greater than 1.

3. Solve the differential equation

$$\frac{d^2y}{dt^2}+2a\frac{dy}{dt}+a^2y=1 \quad (a \text{ constant, } a \neq 0),$$

where $y=\dfrac{dy}{dt}=0$ when $t=0$. Show that the maximum value of $\dfrac{dy}{dt}$ occurs when $t=1/a$.

Two functions $f(t)$, $\phi(t)$ are connected by the equation

$$\frac{d^2f}{dt^2}+2a\frac{df}{dt}+a^2f=\phi.$$

If

$$\phi(t)=\begin{cases}0 & (t<0),\\ 1 & (0<t<t_0),\\ 0 & (t>t_0),\end{cases}$$

and f and $\dfrac{df}{dt}$ are zero for $t\leqslant 0$ and are continuous at the discontinuities of ϕ, show that, for $t>t_0$,

$$f(t)=\frac{e^{-at}}{a^2}\{e^{at_0}(1+at-at_0)-(1+at)\}.$$

4. Verify that, if the function $f(x)$ is continuous for $x>0$, then

$$y=\int_1^x (x-t)\,e^{(x-t)}f(t)\,dt \quad (x>0)$$

is the solution of the equation

$$\frac{d^2y}{dx^2}-2\frac{dy}{dx}+y=f(x),$$

for which $y=0$, $\dfrac{dy}{dx}=0$ when $x=1$.

Find the solution of the equation

$$x\frac{d^2y}{dx^2}-2x\frac{dy}{dx}+xy=e^x,$$

valid for $x>0$, for which $y=0$, $\dfrac{dy}{dx}=0$ when $x=1$.

5. If
$$y=\frac{1}{n}\int_0^x f(t)\sin n(x-t)\,dt,$$
where n is constant, verify that
$$\frac{d^2y}{dx^2}+n^2y=f(x),$$
and that $y=\frac{dy}{dx}=0$ when $x=0$.

Hence, or otherwise, find the solution of the differential equation
$$\frac{d^2y}{dx^2}+y=\sec x,$$
such that $y=a$, $\frac{dy}{dx}=b$ when $x=0$.

6. The electron-density N in a certain ionized gas varies with time t in accordance with the equation
$$\frac{dN}{dt}+\alpha N=I,$$
where $\alpha=\lambda/t$ for $t>0$, and where
$$I=\begin{cases}\mu t(T-t) & t\leqslant T\\ 0 & t\geqslant T,\end{cases}$$
λ, μ, T being real positive constants. If $N=0$ when $t=0$, evaluate N for $t=T$, and show that when $t>T$ the electron-density is
$$\frac{\mu T^{\lambda+3}t^{-\lambda}}{(\lambda+2)(\lambda+3)}.$$

7. The electric current I through a coil of resistance R and inductance L satisfies the differential equation
$$RI+L\frac{dI}{dt}=V,$$
where V is the potential difference between the two ends of the coil. A potential difference $V=a\sin\omega t$ is applied to the coil from the time $t=0$ to the time $t=\pi/\omega$, where a and ω are positive constants. The current is zero at $t=0$, and V is zero after $t=\pi/\omega$. Calculate the current at any time, both before and after $t=\pi/\omega$.

SECTION 2

THE DEFINITION OF FUNCTIONS BY INFINITE SERIES AND INTEGRALS

The functions which we have already met may be summarized briefly. They are the powers of x and combinations of them such as polynomials and rational functions, the trigonometric functions, the logarithmic and exponential functions, and the hyperbolic functions.

The list is substantial, but it forms only a beginning, and many other functions remain for our attention. The work of this section is directed towards the setting-up of certain basic techniques which enable necessary extensions to be made. The question that we pose is less 'What particular functions are there at my disposal?' than 'How can I set about to find such functions when I need them?' The results are all well established, and it is hoped that the presentation will enable the student to see some ways of extending his mathematical vocabulary while at the same time absorbing standard information.

The plan of these four volumes has been to present Calculus in the spirit of Analysis, but without detailed examination of the properties which belong essentially to the latter. It is natural that analytical ideas should become increasingly pressing as we approach the later stages, and it seems wise to insert a section now on convergence and similar topics lest the processes which form our main theme later should be treated on a purely mechanical basis. As in the book as a whole, however, so here also we shall try to clarify the guiding principles rather than to establish the wealth of detail which the serious student of Analysis must always require.

CHAPTER XXIII

THE CONVERGENCE OF SERIES

1. Sequences and series. By a SEQUENCE we mean an *array*

$$s_1, \quad s_2, \quad s_3, \quad \dots$$

of given numbers, real or complex, written in an assigned order; the rth term is denoted by s_r. The terms of the sequence are often formed according to some definite algebraic rule, like the sequence

$$\frac{1}{2}, \quad \frac{1}{9}, \quad \frac{1}{28}, \quad \frac{1}{65}, \quad \frac{1}{126}, \quad \dots,$$

whose rth term is $(r^3+1)^{-1}$. If the sequence terminates, say after n terms, we close the enunciation by inserting the last term; for example

$$s_1, \quad s_2, \quad s_3, \quad \dots, \quad s_n.$$

When there is any room for doubt about the pattern, a 'general term' is written:

$$\frac{1}{3}, \quad \frac{1}{7}, \quad \frac{1}{13}, \quad \frac{1}{21}, \quad \dots, \quad \frac{1}{r^2+r+1}, \quad \dots.$$

A SERIES is a *sum of terms*, often, in work at the present level, formed according to a definite algebraic rule. Familiar examples are the arithmetic series

$$a+(a+d)+(a+2d)+\dots,$$

whose rth term is $a+(r-1)d$, and the geometric series

$$a+at+at^2+\dots,$$

whose rth term is at^{r-1}.

We usually adopt the notation

$$u_1+u_2+u_3+\dots$$

for a series, and denote the sum of the first n terms by the symbol S_n, so that

$$S_n \equiv u_1+u_2+\dots+u_n.$$

The PARTIAL SUMS $\quad S_1, \quad S_2, \quad S_3, \quad \dots$

form the *sequence*

$$u_1, \quad u_1+u_2, \quad u_1+u_2+u_3, \quad \dots.$$

2. Limits and convergence. It may happen that, as r increases, the terms of a sequence

$$s_1, \quad s_2, \quad s_3, \quad \ldots, \quad s_r, \quad \ldots$$

approach some definite LIMIT s. This idea has been considered in Volume I, but is repeated for convenience: *the sequence $s_1, s_2, s_3, \ldots$ approaches a number s as a limit if, given a positive number ϵ, however small,* a number N (depending on ϵ) can be found† so that*

$$|s_r - s| < \epsilon$$

whenever $r > N$. We write $\lim_{r \to \infty} s_r = s$

or $$s_r \to s \quad \text{as} \quad r \to \infty.$$

If, in particular, the sequence

$$S_1, \quad S_2, \quad S_3, \quad \ldots, \quad S_r, \quad \ldots$$

of partial sums of the series

$$u_1 + u_2 + u_3 + \ldots + u_r + \ldots$$

tends to a limit S, we say that the series CONVERGES TO THE SUM S, and we write

$$S = \sum_1^\infty u_r.$$

A series which does not converge is said to DIVERGE.

If a series $$u_1 + u_2 + u_3 + \ldots$$
is convergent, so also is the series found by omitting any given finite number of terms from the beginning. When the number omitted is n, the resulting series is called THE REMAINDER SERIES AFTER n TERMS and (in the case of convergence) is denoted by the symbol R_n, so that

$$R_n \equiv u_{n+1} + u_{n+2} + u_{n+3} + \ldots.$$

Thus $$S \equiv (u_1 + u_2 + \ldots + u_n) + (u_{n+1} + u_{n+2} + \ldots)$$

$$\equiv S_n + R_n.$$

* We shall contract this phrase to 'given ϵ'.

† We shall denote this dependence by the notation

$$N \equiv N(\epsilon),$$

with similar alternatives.

The sum of the first p terms of the remainder series

$$u_{n+1}+u_{n+2}+u_{n+3}+\dots$$

(whether convergent or not), is denoted by the symbol $R_{n,p}$, so that

$$R_{n,p}\equiv u_{n+1}+u_{n+2}+\dots+u_{n+p}.$$

Consideration of the values assumed by R_n for varying values of n leads to the important result that, *if the given series is convergent, then*

$$\lim_{n\to\infty} R_n=0.$$

For, given ϵ, we can find a number $N(\epsilon)$ such that

$$|S-S_n|<\epsilon$$

for all n exceeding N. Hence

$$|R_n|<\epsilon.$$

3. The addition of convergent series. When two series

$$u_1+u_2+u_3+\dots,$$

$$v_1+v_2+v_3+\dots$$

are given, we often wish to add them or to perform other similar processes, such as, say, subtraction, which is merely addition with the sign changed. The theorem which follows justifies the process of term-by-term addition:

If the two sequences U_n, V_n converge to values U, V respectively, then the sequence U_n+V_n converges, and its sum is $U+V$.

Given ϵ, there is a number $M(\epsilon)$ such that

$$|U-U_n|<\tfrac{1}{2}\epsilon$$

whenever $n>M$; and a number $N(\epsilon)$ such that

$$|V-V_n|<\tfrac{1}{2}\epsilon$$

whenever $n>N$. If K is a number greater than both M and N, then, whenever $n>K$, we have the inequality

$$\begin{aligned}|(U+V)-(U_n+V_n)| &\leqslant |U-U_n|+|V-V_n|\\ &<\epsilon.\end{aligned}$$

Hence, by definition of convergence, the sequence (U_n+V_n) converges, and its sum is $U+V$.

In particular, the theorem is true when U_n, V_n are the partial sums of the series

$$u_1 + u_2 + u_3 + \ldots,$$

$$v_1 + v_2 + v_3 + \ldots.$$

That is, *if the two series converge, to sums U and V respectively, then the series*

$$(u_1 + v_1) + (u_2 + v_2) + (u_3 + v_3) + \ldots$$

also converges, to sum $U + V$.

As a direct corollary of the last result we prove the theorem:

If a series

$$u_1 + u_2 + u_3 + \ldots$$

is convergent, then $u_n \to 0$ as $n \to \infty$.

Since the series converges, then, denoting the partial sums by $U_1, U_2, U_3, \ldots$ and their limit by U, we have the relation

$$U_n \to U.$$

It is equally true that $\qquad U_{n-1} \to U.$

But the limit of the difference $U_n - U_{n-1}$ is the difference of the limits $U - U$, so that

$$u_n \to 0.$$

It may be helpful to add the warning that the convergence of a sum does *not* imply the convergence of its constituents. For example, the series

$$(1-1) + (1-1) + (1-1) + \ldots$$

is convergent, but neither of the series

$$1 + 1 + 1 \ldots, \quad -1 - 1 - 1 - \ldots$$

converges. Again, the series

$$\left(\frac{1}{2} + \frac{1}{3}\right) + \left(\frac{1}{2^2} + \frac{1}{3^2}\right) + \left(\frac{1}{2^3} + \frac{1}{3^3}\right) + \left(\frac{1}{2^4} + \frac{1}{3^4}\right) + \ldots$$

is convergent, so the series

$$\left\{\left(\frac{1}{2} + 1\right) + \left(\frac{1}{3} - 1\right)\right\} + \left\{\left(\frac{1}{2^2} + 1\right) + \left(\frac{1}{3^2} - 1\right)\right\} + \left\{\left(\frac{1}{2^3} + 1\right) + \left(\frac{1}{3^3} - 1\right)\right\} + \ldots$$

is convergent also. But neither of the series

$$\left(\frac{1}{2} + 1\right) + \left(\frac{1}{2^2} + 1\right) + \left(\frac{1}{2^3} + 1\right) + \ldots,$$

$$\left(\frac{1}{3} - 1\right) + \left(\frac{1}{3^2} - 1\right) + \left(\frac{1}{3^3} - 1\right) + \ldots$$

converges.

4. The general principle of convergence. The test

$$|R_n| < \epsilon$$

just enunciated is very useful in theoretical applications, but it suffers in practice from the difficulty that we often cannot foresee what the limit S is likely to be, so that the expression $|S - S_n|$ cannot be written down explicitly. There is, however, a principle of very wide application which provides a similar technique with, in many cases, greater ease. This is the GENERAL PRINCIPLE OF CONVERGENCE, which asserts that a NECESSARY AND SUFFICIENT *condition for the convergence of the sequence*

$$s_1, \quad s_2, \quad s_3, \quad \ldots, \quad s_r, \quad \ldots$$

is that, given ϵ, *there exist a number* $N(\epsilon)$ *such that*

$$|s_{n+p} - s_n| < \epsilon$$

for ALL *positive integers* p *and all* $n > N$.

The proof of *sufficiency* involves ideas with which we do not deal. For *necessity*, we proceed as follows:

We are told that the sequence has a limit s. Hence, given ϵ, we can find a number $N(\epsilon)$ sufficiently large to ensure that

$$|s_{n+p} - s| < \tfrac{1}{2}\epsilon, \quad |s_n - s| < \tfrac{1}{2}\epsilon$$

whenever $n+p, n$ exceed N. Hence

$$\begin{aligned} |s_{n+p} - s_n| &= |(s_{n+p} - s) - (s_n - s)| \\ &\leqslant |s_{n+p} - s| + |s_n - s| \\ &< \epsilon, \end{aligned}$$

as required.

When the sequence $\{s_n\}$ is composed of the partial sums $\{S_n\}$ of a series

$$u_1 + u_2 + u_3 + \ldots,$$

the condition appearing in the general principle may be expressed in the alternative forms:

$$\text{(i)} \quad |S_{n+p} - S_n| < \epsilon,$$

or

$$\text{(ii)} \quad |R_{n,p}| < \epsilon,$$

or

$$\text{(iii)} \quad |u_{n+1} + u_{n+2} + \ldots + u_{n+p}| < \epsilon,$$

where n, $n+p$ are, in each case, numbers both exceeding N.

Note as a Corollary the result established earlier by a more direct method, that *it is* NECESSARY *for the convergence of the series*

$$u_1+u_2+\ldots+u_n+\ldots$$

that

$$\lim_{n\to\infty} u_n=0.$$

The condition $u_n\to 0$ is NOT sufficient to ensure convergence. For example, the series

$$1+\frac{1}{\sqrt{2}}+\frac{1}{\sqrt{3}}+\ldots+\frac{1}{\sqrt{n}}+\ldots$$

diverges although $\lim_{n\to\infty}(1/\sqrt{n})\to 0$; for the sum of the first n terms exceeds n times the least of them, so that

$$S_n > n\times\left(\frac{1}{\sqrt{n}}\right)=\sqrt{n} \quad (n>1).$$

Hence

$$S_n\to\infty.$$

We now give a short discussion of some of the simpler properties of real series whose terms are all positive. They play an important part in the more general theory, to which we return later.

5. Real series with positive terms. We require first a basic property whose detailed proof is beyond the scope of this Volume, but the reader will have little difficulty in convincing himself intuitively of its truth. Informally stated, it is that a sequence which increases steadily must *either* tend to infinity *or* to a finite limit—the point being that the alternative of OSCILLATION is prohibited by the steady increase. We therefore assume the truth of the following theorem:

THE 'BOUNDING' TEST. *If the elements*

$$s_1,\quad s_2,\quad s_3,\quad \ldots,\quad s_r,\quad \ldots$$

of a sequence increase steadily, so that

$$s_1\leqslant s_2\leqslant s_3\leqslant\ldots\leqslant s_r\leqslant\ldots,$$

and if, further, the numbers $s_1, s_2, s_3, \ldots, s_r, \ldots$ *are all bounded (that is, all less than some finite number* K*) then the sequence tends to a (finite) limit.*

In particular, if the sequence consists of the partial sums $S_1, S_2, S_3, \ldots$ of a series

$$u_1 + u_2 + u_3 + \ldots$$

of POSITIVE terms, then *either* S_n increases without bound *or* S_n tends to a limit; it cannot oscillate. Thus *a series of* POSITIVE *terms converges if the partial sums S_n are bounded for all values of n, and diverges to $+\infty$ if the sums S_n are not bounded.*

For example, in the geometric series

$$\frac{1}{2} + \frac{1}{2^2} + \frac{1}{2^3} + \ldots,$$

the sum S_n is given by the formula

$$S_n = 1 - \frac{1}{2^n},$$

so that

$$S_n < 1.$$

Hence the sequence $S_1, S_2, S_3, \ldots$, being increasing and bounded, tends to a limit which is the sum of the series; this is, of course, otherwise familiar in this simple case.

The 'bounding test' leads immediately to a 'comparison test' for convergence which is important both in itself and also because it enables us to formulate other specific tests which are easy to apply.

THE COMPARISON TEST. *If each term of the series of* POSITIVE *terms*

$$u_1 + u_2 + u_3 + \ldots + u_r + \ldots$$

is less than, or equal to, the corresponding term of a series of positive terms

$$b_1 + b_2 + b_3 + \ldots + b_r + \ldots$$

known to be convergent, then the given series is also convergent. (The result is not affected if a *finite* number of corresponding terms is omitted from each series before the test is applied.)

The proof is immediate; for

$$S_n \equiv u_1 + u_2 + \ldots + u_n$$

$$\leqslant b_1 + b_2 + \ldots + b_n,$$

and, since the series $b_1 + b_2 + b_3 + \ldots$ is convergent, its sum is bounded. Hence S_n is increasing and bounded, and therefore convergent.

Note the equivalent TEST FOR DIVERGENCE: *If each term of the series of positive terms*

$$u_1 + u_2 + u_3 + \ldots + u_r + \ldots$$

is greater than, or equal to, the corresponding term of a series of positive terms

$$b_1+b_2+b_3+\ldots+b_r+\ldots$$

known to be divergent, then the given series is also divergent.

6. Some detailed tests for series of positive terms. The following well-known tests will be found useful.

(i) D'ALEMBERT'S TEST. *If the ratio* u_n/u_{n+1} *tends to a limit* l, *and if*

$$l>1,$$

then the series $$u_1+u_2+u_3+\ldots+u_r+\ldots$$
is convergent.

Given any positive number ϵ, we can find a number N such that

$$\left|\frac{u_n}{u_{n+1}}-l\right|<\epsilon$$

for all n greater than N. Since u_{n+1} is positive, we may multiply by it, so that

$$|u_n-lu_{n+1}|<\epsilon u_{n+1};$$

that is, $$(l-\epsilon)\,u_{n+1}<u_n<(l+\epsilon)\,u_{n+1}.$$

In particular, we may take ϵ to be the number $\frac{1}{2}(l-1)$, which is positive since $l>1$, and then write

$$k\equiv l-\epsilon=l-\tfrac{1}{2}(l-1)=\tfrac{1}{2}(l+1),$$

so that $$k>1;$$

then the left-hand inequality, $(l-\epsilon)\,u_{n+1}<u_n$, shows that (for the N corresponding to this value of ϵ)

$$u_n>ku_{n+1}\quad(n>N).$$

Thus $$u_{N+2}<u_{N+1}/k,$$
$$u_{N+3}<u_{N+2}/k<u_{N+1}/k^2,$$
$$u_{N+4}<u_{N+3}/k<u_{N+1}/k^3,$$

and so on. Hence each term of the given series after the Nth is less than the corresponding term of the geometric series

$$u_{N+1}\left(1+\frac{1}{k}+\frac{1}{k^2}+\frac{1}{k^3}+\ldots\right)$$

in which the ratio k^{-1} is less than 1, as we have seen. Hence the series converges.

Note that *the test gives no information if* $l=1$; we cannot then take the step of identifying the number ϵ (>0) with $\frac{1}{2}(l-1)$.

The corresponding TEST FOR DIVERGENCE is:

If u_n/u_{n+1} *tends to a limit* l, *and if*

$$l<1,$$

then the series is divergent.

NOTE. It is actually sufficient for convergence that $u_n/u_{n+1}>l$ for all sufficiently large n. The existence of the limit $\lim_{n\to\infty} u_n/u_{n+1}$ is convenient for calculation, but not essential. The proof is easily constructed.

(ii) CAUCHY'S TEST. *To prove that, if* $u_n^{1/n}$ *tends to a limit* l, *and if*

$$l<1,$$

then the series $$u_1+u_2+u_3+\ldots$$
is convergent.

Given any number ϵ, we can find a number N such that

$$|u_n^{1/n}-l|<\epsilon$$

for all n greater than N. Thus

$$l-\epsilon<u_n^{1/n}<l+\epsilon.$$

If, in particular, we take $\epsilon=\frac{1}{2}(1-l)$, which is positive since $l<1$, and write

$$k\equiv l+\epsilon=\tfrac{1}{2}(1+l),$$

so that $$k<1,$$

then the right-hand inequality gives (for the N corresponding to this value of ϵ)

$$u_n^{1/n}<k \quad (n>N),$$

or $$u_n<k^n.$$

The series therefore converges by comparison with the geometric series.

We again have no information if $l=1$, since the step $\epsilon=\frac{1}{2}(1-l)$ is then excluded.

The corresponding TEST FOR DIVERGENCE is:

If $u_n^{1/n}$ *tends to a limit* l, *and if*

$$l>1,$$

then the series $$u_1+u_2+u_3+\ldots$$
is divergent.

NOTE. The simpler form, that the series converges if

$$u_n^{1/n} < l \quad (l < 1),$$

for all sufficiently large n, is very easily obtained; for the term u_n is then less than the corresponding term l^n of the convergent geometrical series

$$l + l^2 + l^3 + \ldots,$$

and the result follows from the Comparison Test.

(iii) There are many tests, not considered here, which can be applied to series where the limits u_n/u_{n+1} and $u_n^{1/n}$ have the excluded value 1. It is, however, useful to have a test which often establishes at once the negative result that a series of positive terms is *not* convergent:

If

$$u_1 + u_2 + u_3 + \ldots$$

is a series of positive decreasing terms, and if the series does converge, then

$$\lim_{n \to \infty} n u_n = 0.$$

If the series converges, we can, given any positive number ϵ, however small, find a number N (depending on ϵ) such that the remainder after N terms (p. 78) is less than $\frac{1}{2}\epsilon$; that is, such that

$$R_n \equiv u_{N+1} + u_{N+2} + \ldots < \tfrac{1}{2}\epsilon.$$

Then, *a fortiori*,

$$u_{N+1} + u_{N+2} + \ldots + u_n < \tfrac{1}{2}\epsilon \quad (n > N);$$

and, since the terms are decreasing, each of the $n - N$ terms on the left exceeds u_n, so that the last inequality gives

$$(n - N)\, u_n < \tfrac{1}{2}\epsilon,$$

or

$$n u_n < \tfrac{1}{2}\epsilon + N u_n.$$

Moreover, it is necessary (p. 80) that in a convergent series

$$u_n \to 0,$$

and so n may be taken sufficiently large to ensure the inequality

$$u_n < \tfrac{1}{2}\epsilon/N,$$

or

$$N u_n < \tfrac{1}{2}\epsilon.$$

Then

$$n u_n < \tfrac{1}{2}\epsilon + N u_n$$

$$< \epsilon,$$

so that $$\lim_{n\to\infty} nu_n = 0.$$

For example, the harmonic series
$$1+\tfrac{1}{2}+\tfrac{1}{3}+\ldots$$
has decreasing terms, and
$$nu_n = n(1/n) = 1 \nrightarrow 0.$$
Hence this series does not converge.

Note that the simple requirement '$u_n \to 0$ for a convergent series' will of itself sometimes settle that a series cannot converge. For example, in the series
$$1+\tfrac{1}{2}+\tfrac{2}{3}+\tfrac{3}{4}+\tfrac{4}{5}+\ldots,$$
we have
$$u_n = \frac{n-1}{n} \to 1,$$
so that the series does not converge.

(iv) RAABE'S TEST. We conclude with one illustrative test for certain series in which $u_n/u_{n+1} \to 1$:

If the limit
$$\lim_{n\to\infty}\left\{n\left(\frac{u_n}{u_{n+1}}-1\right)\right\}$$
exists and has value l, *where* $l>1$,

then the series $$u_1+u_2+u_3+\ldots+u_r+\ldots$$
converges.

Given any positive number ϵ, we can find a number N, depending on ϵ, such that
$$l-\epsilon < n\left(\frac{u_n}{u_{n+1}}-1\right) < l+\epsilon$$
whenever $n > N$. If we multiply by u_{n+1}, which is positive, the left-hand inequality gives
$$(l-\epsilon)\,u_{n+1} < nu_n - nu_{n+1},$$
or
$$(l-1-\epsilon)\,u_{n+1} < nu_n - (n+1)\,u_{n+1}.$$
Giving to n the values $N+1, N+2, N+3, \ldots$ in succession, we have the inequalities
$$(l-1-\epsilon)\,u_{N+2} < (N+1)\,u_{N+1} - (N+2)\,u_{N+2},$$
$$(l-1-\epsilon)\,u_{N+3} < (N+2)\,u_{N+2} - (N+3)\,u_{N+3},$$
$$\ldots\ldots\ldots\ldots\ldots\ldots\ldots\ldots\ldots\ldots\ldots\ldots\ldots\ldots$$
$$(l-1-\epsilon)\,u_M < (M-1)\,u_{M-1} - Mu_M$$

for any number M greater than N. Adding, and writing

$$S_n \equiv u_1 + u_2 + \ldots + u_n$$

for the nth partial sum, we reach the formula

$$(l-1-\epsilon)(S_M - S_{N+1}) < (N+1)\,u_{N+1} - Mu_M,$$

or, since Mu_M is positive,

$$(l-1-\epsilon)(S_M - S_{N+1}) < (N+1)\,u_{N+1}.$$

Now we are given that $l > 1$, and so we can choose ϵ so that $l-1-\epsilon > 0$. Then

$$S_M < S_{N+1} + \frac{(N+1)\,u_{N+1}}{l-1-\epsilon}.$$

For any given ϵ, the right-hand side is determinate, since ϵ determines N. Hence the sequence of which S_M is a typical element $(M > N)$ is (i) increasing, since $u_1, u_2, u_3, \ldots$ are all positive, and (ii) bounded, by what we have just proved. Hence S_M has a limit S to which the given series converges.

We have also the *test for divergence*:

If the limit

$$\lim_{n\to\infty} \left\{ n\left(\frac{u_n}{u_{n+1}} - 1\right)\right\}$$

exists and has value l, *where* $l < 1$,
then the series diverges.

EXAMPLES I

Test for convergence the series with the following nth terms:

1. $n/2^n$.
2. $n^2/3^n$.
3. $(\frac{1}{2}n)^n$.
4. $(3/n)^n$.
5. $(2n+1)/2^n$.
6. $n!/n^2$.
7. $(n+2)^2/(2n)!$.
8. $(n!)^{-n}$.
9. $\left(\dfrac{n+2}{3n+4}\right)^{5n}$.
10. $\dfrac{n+2}{3n+4}$.
11. $\dfrac{1}{n^3}$.
12. $\dfrac{2n+1}{n^2}$.
13. $\dfrac{n-1}{n^3+1}$.
14. $\left(\dfrac{n+3}{n-3}\right)^2$.
15. $\dfrac{n!}{(2n)!}$.

7. More general series; absolute convergence. Experience shows that the mere convergence of a series does not always suffice for the manipulations required of it. For example, it is fairly easy to prove that a convergent series of positive terms has a definite sum independently of the order in which its terms are taken. But this is not necessarily true when some of the terms are positive and some negative; thus the two series

$$1-\tfrac{1}{2}+\tfrac{1}{3}-\tfrac{1}{4}+\tfrac{1}{5}-\tfrac{1}{6}+\tfrac{1}{7}-\tfrac{1}{8}+\dots,$$

$$1+\tfrac{1}{3}+\tfrac{1}{5}-\tfrac{1}{2}+\tfrac{1}{7}+\tfrac{1}{9}+\tfrac{1}{11}-\tfrac{1}{4}+\dots$$

(with three positives before each negative in the second) have quite different sums.

It is found necessary to consider, as a sub-class of the convergent series, those for which the series of the *moduli* of the terms are also convergent. We therefore make the definition:

The series

$$u_1+u_2+u_3+\dots$$

is said to CONVERGE ABSOLUTELY *if the series of moduli*

$$|u_1|+|u_2|+|u_3|+\dots$$

is convergent.

The concept of absolute convergence enables us to make use of the tests already found for series of positive terms.

8. The convergence of an absolutely convergent series. It is possible for a series to be convergent although *not* absolutely; for example, the two series quoted in §7 are convergent, but the series of their moduli (absolute values) are not. On the other hand, we prove almost immediately that a series which is 'absolutely convergent' does actually converge.

Let it be given that the series

$$u_1+u_2+u_3+\dots,$$

whose terms we suppose to be real, converges absolutely, so that the series of moduli

$$|u_1|+|u_2|+|u_3|+\dots$$

converges to a value T. We prove that *the series formed by taking the positive terms (in order, as they come) converges to a value P, that the series formed by taking the negative terms converges to a (negative) value $-Q$, and that the given series converges to the value $P-Q$.*

Write

$$p_n\equiv\tfrac{1}{2}(|u_n|+u_n),$$

$$q_n\equiv\tfrac{1}{2}(|u_n|-u_n).$$

Then $p_n \equiv u_n$ when u_n is positive, and $p_n \equiv 0$ when u_n is negative; also $q_n \equiv -u_n$ when u_n is negative, and $q_n \equiv 0$ when u_n is positive. Thus the terms of the two series

$$p_1 + p_2 + p_3 + \ldots,$$

$$q_1 + q_2 + q_3 + \ldots$$

are all positive, and equal to or less than the corresponding terms of the convergent series

$$|u_1| + |u_2| + |u_3| + \ldots.$$

Hence, by the Comparison Test, the two series converge, say to values P and Q.

Moreover, we have established (p. 80) that the series whose terms are the difference of corresponding terms of two series of positive terms is itself convergent, to the difference of the sums. Hence the series

$$(p_1 - q_1) + (p_2 - q_2) + (p_3 - q_3) + \ldots$$

converges to the value $P - Q$.

But

$$p_n - q_n \equiv u_n$$

for all values of n, so that the series

$$u_1 + u_2 + u_3 + \ldots$$

converges to the value $P - Q$.

NOTE. We have confined our attention to real series, but it is also true for a series of *complex* terms

$$u_1 + u_2 + u_3 + \ldots$$

that, if it is absolutely convergent, it is also convergent. This follows easily from the general principle of convergence. For, if we write

$$S_{n,p} \equiv u_{n+1} + u_{n+2} + \ldots + u_{n+p},$$

$$T_{n,p} \equiv |u_{n+1}| + |u_{n+2}| + \ldots + |u_{n+p}|,$$

then

$$\begin{aligned} |S_{n,p}| &\equiv |u_{n+1} + u_{n+2} + \ldots + u_{n+p}| \\ &\leqslant |u_{n+1}| + |u_{n+2}| + \ldots + |u_{n+p}| \\ &\leqslant T_{n,p}. \end{aligned}$$

For absolute convergence, $T_{n,p} \to 0$, so that $|S_{n,p}| \to 0$ also. Hence the series is convergent.

9. Alternating series. Among the series which may converge, though not necessarily absolutely, an important class consists of the ALTERNATING SERIES whose terms are real and alternatively positive and negative in sign; say

$$u_1 - u_2 + u_3 - u_4 + \dots \quad (u_r > 0).$$

We prove that, *if the terms of an alternating series decrease steadily in magnitude, so that*

$$u_1 > u_2 > u_3 > u_4 > \dots,$$

and tend to zero with n, so that

$$\lim_{n\to\infty} u_n = 0,$$

then the series is convergent.

Consider first the sum of an even number of terms,

$$S_{2n} \equiv (u_1 - u_2) + (u_3 - u_4) + \dots + (u_{2n-1} - u_{2n}).$$

Each difference $(u_{2r-1} - u_{2r})$ is positive, and so S_{2n} is (with terms so grouped) an increasing sequence. Also

$$S_{2n} \equiv u_1 - (u_2 - u_3) - (u_4 - u_5) - \dots - (u_{2n-2} - u_{2n-1}) - u_{2n},$$

so that

$$S_{2n} < u_1.$$

Hence the sequence S_{2n}, being (as grouped) increasing and bounded, tends to a limit S.

Moreover,

$$S_{2n+1} = S_{2n} + u_{2n+1},$$

so that

$$\lim_{n\to\infty} S_{2n+1} = \lim_{n\to\infty} S_{2n} + \lim_{n\to\infty} u_{2n+1}$$
$$= S.$$

Thus

$$\lim_{n\to\infty} S_r = S$$

whether r is odd or even, and so the series converges.

NOTE. The condition $\lim_{n\to\infty} u_n = 0$, requisite (p. 80) for all series, is essential to the argument. For example, the terms of the series

$$1 - \tfrac{2}{3} + \tfrac{3}{5} - \tfrac{4}{7} + \tfrac{5}{9} - \tfrac{6}{11} + \dots$$

alternate in sign and decrease steadily in magnitude; but the series does not converge since

$$\lim_{n\to\infty} \frac{n}{2n-1} = \frac{1}{2}.$$

For large values of n the series oscillates by the alternate addition and subtraction of an amount approximately equal to $\frac{1}{2}$.

10. The effect of changing the order of terms. In general it is not possible to alter the order in which the terms of an infinite series appear without at the same time altering its value. For example, if order were irrelevant, we should be able to use the argument

$$\begin{aligned}\log_e 2 &= 1-\tfrac{1}{2}+\tfrac{1}{3}-\tfrac{1}{4}+\tfrac{1}{5}-\tfrac{1}{6}+\dots\\ &= 1-\tfrac{1}{2}-\tfrac{1}{4}+\tfrac{1}{3}-\tfrac{1}{6}-\tfrac{1}{8}+\tfrac{1}{5}-\tfrac{1}{10}-\tfrac{1}{12}+\dots\\ &= (1-\tfrac{1}{2})-\tfrac{1}{4}+(\tfrac{1}{3}-\tfrac{1}{6})-\tfrac{1}{8}+(\tfrac{1}{5}-\tfrac{1}{10})-\tfrac{1}{12}+\dots\\ &= \tfrac{1}{2}-\tfrac{1}{4}+\tfrac{1}{6}-\tfrac{1}{8}+\tfrac{1}{10}-\tfrac{1}{12}+\dots\\ &= \tfrac{1}{2}(1-\tfrac{1}{2}+\tfrac{1}{3}-\tfrac{1}{4}+\tfrac{1}{5}-\dots)\\ &= \tfrac{1}{2}\log_e 2,\end{aligned}$$

so that
$$\log_e 2 = 0,$$
or
$$2 = 1.$$

We prove, however, that, *if a series converges* ABSOLUTELY, *then its sum is unaltered by any change in the order of its terms.*

Take first a convergent series of positive terms

$$S \equiv u_1+u_2+u_3+\dots,$$

and suppose that it is rearranged so that the terms appear in a different order, forming a series which we write in the form

$$v_1+v_2+v_3+\dots.$$

Let T_m be the sum of the first m terms of the rearranged series, and suppose that n terms of the given series must be taken before all the terms of T_m have been included. Then

$$T_m \leqslant S_n < S,$$

so that the sequence T_m, being increasing and bounded, converges to a limit T, where
$$T \leqslant S.$$

Now that we know the rearranged series to be convergent, we may interchange the parts played by S and T in the argument just used, so that
$$S \leqslant T.$$

Hence
$$S = T,$$
as we wished to prove.

Suppose next that the series S contains both positive and negative terms, but that it is given to converge absolutely. Then (p. 89) its positive terms converge to a sum P and its negative terms to a (negative) sum $-Q$, where

$$S=P-Q.$$

Any change in the order of terms does not change the value of the sum P of all the positive terms or (by similar argument) the sum $-Q$ of all the negative terms. Hence the rearranged series converges absolutely, and its sum is still $P-Q$.

EXAMPLES II

Test for convergence the series with the following nth terms:

1. $(-1)^n n^{-2}$.
2. $(-1)^n n^{-n}$.
3. $(-1)^n n^{\frac{1}{2}}$.
4. $\dfrac{(-1)^n}{n!}$.
5. $(-1)^n \log n$.
6. $\dfrac{(-1)^n \log n}{n}$.
7. $(-1)^n \dfrac{n+2}{n+3}$.
8. $(-1)^n \dfrac{n-5}{5n-1}$.
9. $\dfrac{(-1)^n n^2}{(n+1)^4}$.
10. $(-1)^n$.
11. $(-1)^n/\sqrt{n}$.
12. $(-\frac{1}{2})^n$.
13. $(-1)^n e^{-n} \log n$.
14. $(-1)^n n^2 e^{-n}$.
15. $(-1)^n e^{\sqrt{n}}$.
16. $(-1)^n \dfrac{(e^n+1)}{(e^{2n}+1)}$.

11. The integral test. There is a useful theorem for comparing the convergence of an infinite series with that of a closely related infinite integral. Let

$$f(x)$$

be a function, *positive and decreasing for all positive values of* x; write

$$S_r \equiv f(1)+f(2)+\ldots+f(r),$$

$$U_r \equiv \int_1^r f(x)\,dx.$$

Then *the difference* $$g_n \equiv S_n - U_n$$

tends to a (finite) limit as n tends to infinity:

Since $f(x)$ decreases as x increases, we have, for $r = 2, 3, 4, \ldots$, the inequalities

$$f(r-1) \geqslant f(x) \geqslant f(r) \quad (r-1 \leqslant x \leqslant r),$$

so that, on integrating with respect to x between the limits $r-1, r$,

$$f(r-1) \geqslant \int_{r-1}^{r} f(x)\,dx \geqslant f(r).$$

From the left-hand inequality, it follows that

$$f(1) + f(2) + \ldots + f(n-1) \geqslant \int_1^n f(x)\,dx,$$

or $$S_n - f(n) \geqslant U_n,$$

or $$g(n) \geqslant f(n).$$

But we are given that $f(x)$ is always positive, so that

$$g(n) \geqslant 0.$$

Moreover, the right-hand inequality

$$\int_{r-1}^{r} f(x) \geqslant f(r)$$

may be written in the form

$$U_r - U_{r-1} \geqslant S_r - S_{r-1},$$

or $$g_r \leqslant g_{r-1}.$$

Thus $$g_1 \geqslant g_2 \geqslant g_3 \geqslant \ldots.$$

The function g_n therefore decreases with n, but is always positive. Hence it tends to a finite limit, so that

$$\lim_{n \to \infty} (S_n - U_n)$$

exists and is finite.

COROLLARY. *If either the sum*

$$f(1) + f(2) + f(3) + \ldots + f(n)$$

or the integral $$\int_1^n f(x)\,dx$$

converges (or diverges), when $f(x)$ is a positive decreasing function of x, so also does the other.

12. Euler's constant γ. The integral test leads immediately to an important result:

To prove that *the function*

$$1+\frac{1}{2}+\frac{1}{3}+\dots+\frac{1}{n}-\log n$$

tends to a limit γ as n tends to infinity.

In the integral test, take

$$f(x)\equiv\frac{1}{x}\quad(x>0).$$

Then the conditions (i) positive, (ii) decreasing are both satisfied, so that the limit of the function

$$1+\frac{1}{2}+\frac{1}{3}+\dots+\frac{1}{n}-\int_1^n\frac{dx}{x}$$

exists and is finite; that is, the limit

$$\lim_{n\to\infty}\left(1+\frac{1}{2}+\frac{1}{3}+\dots+\frac{1}{n}-\log n\right)$$

exists, having a value which we may call γ.

The constant γ is known as EULER'S CONSTANT.

ILLUSTRATION 1. *To find the sum of the series*

$$1+\tfrac{1}{3}-\tfrac{1}{2}+\tfrac{1}{5}+\tfrac{1}{7}-\tfrac{1}{4}+\tfrac{1}{9}+\tfrac{1}{11}-\tfrac{1}{6}+\dots,$$

where two positive terms from the series

$$1+\tfrac{1}{3}+\tfrac{1}{5}+\tfrac{1}{7}+\dots$$

are followed by one negative term from the series

$$-\tfrac{1}{2}-\tfrac{1}{4}-\tfrac{1}{6}-\dots.$$

The sum S_{3n} of the first $3n$ terms may be written in the form

$$\left\{\left(1+\frac{1}{2}+\dots+\frac{1}{4n}\right)-\left(\frac{1}{2}+\frac{1}{4}+\dots+\frac{1}{4n}\right)\right\}-\left(\frac{1}{2}+\frac{1}{4}+\dots+\frac{1}{2n}\right).$$

Write
$$1+\frac{1}{2}+\frac{1}{3}+\dots+\frac{1}{r}-\log r\equiv\gamma+\epsilon_r,$$

where
$$\lim_{r\to\infty}\epsilon_r=0.$$

Then

$$S_{3n} = \{(\gamma + \log 4n + \epsilon_{4n}) - \tfrac{1}{2}(\gamma + \log 2n + \epsilon_{2n})\} - \tfrac{1}{2}(\gamma + \log n + \epsilon_n)$$
$$= \log 4n - \tfrac{1}{2}\log 2n - \tfrac{1}{2}\log n + \epsilon_{4n} - \tfrac{1}{2}\epsilon_{2n} - \tfrac{1}{2}\epsilon_n$$
$$= \tfrac{1}{2}\log\left(\frac{16n^2}{2n^2}\right) + \epsilon_{4n} - \tfrac{1}{2}\epsilon_{2n} - \tfrac{1}{2}\epsilon_n.$$

Hence

$$\lim_{n\to\infty} S_{3n} = \tfrac{1}{2}\log 8.$$

Since also the individual terms of the series all tend to zero, it follows that S_{3n+1}, S_{3n+2} both tend to the same sum, so that the series converges, to the value $\tfrac{1}{2}\log 8$.

13. The series Σn^{-s}. A very useful test for comparison purposes is afforded by the series

$$1 + \frac{1}{2^s} + \frac{1}{3^s} + \frac{1}{4^s} + \dots \quad (s \text{ real}).$$

We prove that *the series converges when $s > 1$ and diverges when $s \leqslant 1$.*

The work of the preceding paragraph (and also of p. 87) establishes divergence for $s = 1$, since it is well known (Volume II, p. 3) that

$$\log n \to \infty.$$

This carries with it divergence for $s < 1$, since then

$$\frac{1}{n^s} > \frac{1}{n},$$

so that, for any value of N,

$$\sum_{n=1}^{N} \frac{1}{n^s} > \sum_{n=1}^{N} \frac{1}{n}.$$

For $s > 1$ we apply the integral test (p. 93). The integral

$$\int_1^n \frac{dx}{x^s}$$

has the value $$\left[\frac{-1}{(s-1)x^{s-1}}\right]_1^n = \frac{1}{s-1} - \frac{1}{(s-1)n^{s-1}},$$

so that, for $s > 1$, $$\int_1^n \frac{dx}{x^s} \to \frac{1}{s-1}.$$

Since the function x^{-s} is positive and decreasing for positive x, the conditions of the integral test are satisfied, so that the series converges with the integral for $s > 1$.

14. The sequence $\left(1+\frac{x}{n}\right)^n$. Another particular result which is found useful in applications is that *the sequence*

$$S_n \equiv \left(1+\frac{x}{n}\right)^n$$

tends, for real values of x, to the limit

$$e^x$$

as n tends to infinity.

Taking logarithms, we have the relation

$$\log_e S_n = n \log_e \left(1+\frac{x}{n}\right).$$

If we write

$$y = 1/n,$$

the relation becomes

$$\log_e S_n = \frac{\log_e(1+xy)}{y},$$

and so

$$\lim_{n\to\infty} \log_e S_n = \lim_{y\to 0} \frac{\log_e(1+xy)}{y}$$

for given x. But it is easy to prove that

$$\lim_{y\to 0} \frac{\log_e(1+xy)}{y} = x,$$

so that

$$\lim_{n\to\infty} \log_e S_n = x.$$

Since, then,

$$\log_e S_n \to x,$$

it follows, from the continuity of the function e^x, that

$$S_n \equiv e^{\log_e S_n} \to e^x.$$

ILLUSTRATION 2. *To express* $\log(1+z)$ *in a series of ascending powers of z, where z is* COMPLEX.

Suppose that z is expressed in modulus-argument form,

$$z = re^{i\theta}.$$

Taking the hint from the definition for real values of z, we consider the integral

$$I \equiv \int_0^r \frac{e^{i\theta}\,dt}{1+te^{i\theta}} \quad (t \text{ real}),$$

and express it first in terms of its real and imaginary parts; thus

$$I = \int_0^r \frac{e^{i\theta}(1+te^{-i\theta})}{(1+te^{i\theta})(1+te^{-i\theta})}\,dt$$

$$= \int_0^r \frac{(\cos\theta + t) + i\sin\theta}{1+2t\cos\theta+t^2}\,dt$$

$$= \tfrac{1}{2}\log(1+2r\cos\theta+r^2) + i\tan^{-1}\left(\frac{r\sin\theta}{1+r\cos\theta}\right)$$

for appropriate choice of the inverse tangent.

Alternatively, we have by expansion the identity

$$I = \int_0^r \left\{e^{i\theta} - te^{2i\theta} + t^2e^{3i\theta} - \ldots + (-)^{m-1}t^{m-1}e^{mi\theta} + \frac{(-)^m t^m e^{(m+1)i\theta}}{1+te^{i\theta}}\right\}dt$$

$$= re^{i\theta} - \tfrac{1}{2}r^2e^{2i\theta} + \ldots + (-)^{m-1}\frac{1}{m}r^m e^{mi\theta} + R_m$$

$$= z - \tfrac{1}{2}z^2 + \ldots + (-)^{m-1}\frac{1}{m}z^m + R_m,$$

where
$$R_m \equiv \int_0^r \frac{(-)^m t^m e^{(m+1)i\theta}}{1+te^{i\theta}}\,dt,$$

so that
$$|R_m| \leqslant \int_0^r \frac{t^m\,dt}{|1+te^{i\theta}|}.$$

Now the real and imaginary parts of the denominator are

$$1+t\cos\theta, \quad t\sin\theta,$$

and they cannot *both* vanish except when

$$\theta = (2k+1)\pi, \quad t = 1$$

(remembering that t is positive, since r is). Thus $|1+te^{i\theta}|$ is definitely positive (not zero) except for $\theta = (2k+1)\pi$; say

$$|1+te^{i\theta}| > \delta \quad (\theta \neq (2k+1)\pi).$$

Hence
$$|R_m| < \frac{r^{m+1}}{(m+1)\delta}.$$

If, therefore, we make the restriction

$$r \leqslant 1,$$

then
$$|R_m| \to 0.$$

Hence, if $|z| \leqslant 1$, $z \neq -1$, then

$$z - \tfrac{1}{2}z^2 + \tfrac{1}{3}z^3 - \ldots = \tfrac{1}{2}\log(1+2r\cos\theta+r^2) + i\tan^{-1}\left(\frac{r\sin\theta}{1+r\cos\theta}\right),$$

where $z = re^{i\theta}$.

Finally, suppose that

$$\log(1+z)=u+iv,$$

so that $$1+r\cos\theta+ir\sin\theta=e^u(\cos v+i\sin v).$$

Then $$1+r\cos\theta=e^u\cos v,$$

$$r\sin\theta=e^u\sin v,$$

and so $$1+2r\cos\theta+r^2=e^{2u},$$

$$\frac{r\sin\theta}{1+r\cos\theta}=\tan v,$$

or $$u=\tfrac{1}{2}\log(1+2r\cos\theta+r^2),$$

$$v=\tan^{-1}\left(\frac{r\sin\theta}{1+r\cos\theta}\right),$$

for appropriate choice of the inverse tangent. Thus

$$z-\tfrac{1}{2}z^2+\tfrac{1}{3}z^3-\ldots=u+iv \quad (|z|\leqslant 1,\ z\neq -1)$$

$$=\log(1+z).$$

The ambiguity in the interpretation of the inverse tangents is resolved by noting that, when $z=0$, the sum of the series is zero, so that that value of the logarithm must be taken whose argument lies between $-\pi$ and π. The value of θ may be restricted to the interval $(-\pi,\pi)$, and the inverse tangents are taken to vanish with θ.

NOTE. This seems at first sight a long way round for a formula which is identical in form to that obtained for real z; but the treatment for real variable does not apply for complex. The ambiguity in the value of $\log(1+z)$ for complex numbers (reflected in the arrival of the inverse tangent in our proof) is in itself a warning that complications might be expected.

REVISION EXAMPLES XVIII

1. Find the sum to n terms of each of the series

$$\tfrac{1}{2}+\tfrac{1}{4}+\tfrac{1}{8}+\tfrac{1}{16}+\ldots,$$

$$1^2+2^2+3^2+4^2+\ldots.$$

Deduce which of these series converges, and the value of its sum to infinity.

2. Find the sum of the first n terms of the series

$$2-\tfrac{2}{3}+\tfrac{2}{9}-\tfrac{2}{27}+\tfrac{2}{81}-\dots.$$

Determine whether the series is convergent and, if so, find its sum to infinity.

3. Sum to n terms each of the series

$$1+2+3+4+\dots,$$

$$\frac{1}{1.2}+\frac{1}{2.3}+\frac{1}{3.4}+\dots.$$

Deduce which of these series converges, and the value of its sum to infinity.

4. Show that the sum of the first n terms of the series

$$\frac{1}{1.3}+\frac{1}{3.5}+\frac{1}{5.7}+\frac{1}{7.9}+\dots$$

is

$$\frac{1}{2}\left(1-\frac{1}{2n+1}\right).$$

Hence show that the infinite series converges, and find its sum.

5. Find the sum to n terms of each of the series

$$\frac{1}{1.4}+\frac{1}{4.7}+\dots+\frac{1}{(3r-2)(3r+1)}+\dots,$$

$$1.2+2.3+\dots+r(r+1)+\dots.$$

Deduce which of these series converges, and the value of its sum to infinity.

6. Examine whether the series

$$\frac{1}{3}-\frac{2}{5}+\frac{3}{7}-\dots+(-1)^{n-1}\frac{n}{2n+1}+\dots$$

converges or not.

7. Find numbers A, B such that

$$\frac{1}{x^2-4}\equiv\frac{A}{x-2}+\frac{B}{x+2}.$$

Hence, or otherwise, find (i) the sum to n terms, (ii) the sum to infinity of the series

$$\frac{1}{3^2-4}+\frac{1}{4^2-4}+\frac{1}{5^2-4}+\dots.$$

8. Prove that the sum to n terms of the series

$$\frac{1}{2.3.4}+\frac{1}{3.4.5}+\frac{1}{4.5.6}+\dots$$

is

$$\frac{1}{12}-\frac{1}{2(n+2)(n+3)},$$

and deduce that the sum to infinity is $\frac{1}{12}$.

9. Find the sum of the first n terms of each of the series

$$\text{(i)}\quad \frac{1}{1.2}+\frac{1}{2.3}+\frac{1}{3.4}+\dots,$$

$$\text{(ii)}\quad 1.2.5+2.3.6+3.4.7+\dots.$$

Examine the convergence as n tends to infinity.

10. Sum to infinity the series

$$\sin A+\tfrac{1}{2}\sin 2A+\tfrac{1}{4}\sin 3A+\tfrac{1}{8}\sin 4A+\dots.$$

11. Sum to infinity each of the series

$$C\equiv\cos\theta\cos\theta+\cos^2\theta\,\cos 2\theta+\cos^3\theta\,\cos 3\theta+\cos^4\theta\,\cos 4\theta+\dots$$

and

$$S\equiv\cos\theta\,\sin\theta+\cos^2\theta\,\sin 2\theta+\cos^3\theta\,\sin 3\theta+\cos^4\theta\,\sin 4\theta+\dots,$$

and state the values of θ, if any, for which the series do *not* converge.

12. Find the sum of n terms of the series whose rth term is

$$\frac{2r-1}{r(r+1)(r+2)},$$

and find whether the sum to infinity exists.

13. Prove that, if $n>1$,

$$\frac{1}{1.2.3}+\frac{1}{2.3.4}+\dots+\frac{1}{(n-1)n(n+1)}=\frac{(n-1)(n+2)}{4n(n+1)}.$$

Show that this series has a sum to infinity, and find roughly how many terms of the series must be taken to give a sum differing from the sum to infinity by not more than one part in a million.

14. Show that the series

$$\frac{1}{1+a}+\frac{1}{1+a^2}+\frac{1}{1+a^3}+\frac{1}{1+a^4}+\dots,$$

where $a>0$, is convergent if $a>1$ and divergent if $a\leqslant 1$.

15. Expose any fallacy in the argument:

For the series Σu_n, where $u_n \equiv 1/n$, the ratio u_n/u_{n+1} always exceeds 1, and so the series converges.

Would the argument have held for the series with $u_n \equiv 1/n^2$?

16. Two numbers a_1, b_1 are given, where $a_1 > b_1 > 0$. The pair of numbers a_n, b_n are defined for $n \geqslant 2$ as the arithmetic and harmonic means respectively of the pair a_{n-1}, b_{n-1}. Prove that

(i) $a_n b_n = a_1 b_1$,

(ii) the sequence a_n decreases and the sequence b_n increases,

(iii) as n tends to infinity, a_n and b_n both tend to the limit $\sqrt{(a_1 b_1)}$.

17. Prove that, if

$$u_n \equiv \frac{1.3.5.\ldots.(2n-1)}{2.4.6.\ldots.(2n)},$$

then nu_n^2 is an increasing sequence and $(n+\frac{1}{2})u_n^2$ a decreasing sequence.

Deduce that nu_n^2 tends to a finite positive limit as $n \to \infty$, stating without proof any general theorem on sequences to which you appeal.

18. Two infinite sequences $\{a_n\}$, $\{b_n\}$ are defined in terms of two given numbers a, b $(a > b > 0)$ as follows:

(i) $a_0 = a$, $b_0 = b$;

(ii) a_n is the arithmetic mean, and b_n is the geometric mean, of a_{n-1} and b_{n-1} $(n \geqslant 1)$.

Prove that a_n and b_n tend to a common limit as $n \to \infty$.

19. Prove that, if $|x| < 1$, then

$$\lim_{n\to\infty} x^n = 0.$$

Prove that, if

$$\lim_{n\to\infty} \frac{a_{n+1}}{a_n} = l,$$

and if $0 \leqslant l < 1$, then

$$\lim_{n\to\infty} a_n = 0.$$

Deduce the limit as n tends to infinity of

$$\frac{n^3}{2^n}, \quad \frac{10^n}{n!}, \quad \frac{3.5.7.\ldots.(2n+1)}{1.4.7.\ldots.(3n-2)}.$$

20. The terms of a series Σu_n are all positive. State which of the following conditions is, or are, sufficient for the convergence of the series:

(i) $u_{n+1}/u_n < 1$ for all n;

(ii) there exists a number N, and a number $k \leqslant 1$, such that $u_{n+1}/u_n < k$ for all $n > N$;

(iii) there exists a number N, and a number $k < 1$, such that $u_{n+1}/u_n \leqslant k$ for all $n > N$;

(iv) $\lim\limits_{n\to\infty} (u_{n+1}/u_n) \leqslant 1$;

(v) $\lim\limits_{n\to\infty} (u_{n+1}/u_n) < 1$.

For each of the conditions which you state are sufficient for convergence, explain whether it is also necessary, and if not, give an example to illustrate that it is not.

Show, by illustrative examples or otherwise, that the other conditions are not correct as sufficient conditions for the convergence of the series.

21. Prove that, if $x_n \to l$ as $n \to \infty$,

and if $$l \leqslant y_n \leqslant x_n$$

for all n, then $$y_n \to l \quad \text{as} \quad n \to \infty.$$

Prove that, if $\alpha > 0$, then

$$\left(1+\frac{1}{n^{1+\alpha}}\right)^n < 1+\frac{1}{n^\alpha}\left(1-\frac{1}{2n^\alpha}\right)^{-1}$$

for all positive integral n. Hence show that

$$\left(1+\frac{1}{n^{1+\alpha}}\right)^n \to 1 \quad \text{as} \quad n \to \infty.$$

22. Prove that the series $$\sum_1^\infty \frac{1}{n^s}$$

converges if $s > 1$ and diverges if $s \leqslant 1$.

Writing $c_n \equiv n^{-s}$, prove that

$$\lim_{n\to\infty} n\left(\frac{c_n}{c_{n+1}} - 1\right) = s.$$

Discuss the convergence of the (real) series whose nth term is

$$\frac{(a+1)(a+2)\dots(a+n)}{n!}.$$

23. If
$$a_{n+1}=pa_n+qa_{n-1},$$
where p, q are positive, and if α, β are the positive and negative roots respectively of the equation
$$x^2=px+q,$$
prove that, if
$$u_n\equiv a_n/\alpha^n,$$
then
$$u_{n+1}-u_n=(u_n-u_{n-1})(\beta/\alpha),$$
and deduce that the sequence (u_n) converges to the value
$$\frac{a_2-\beta a_1}{\alpha(\alpha-\beta)}.$$

24. Prove that, if the sequence (u_n) decreases steadily as n tends to infinity, and if
$$\sum_{n=1}^{\infty} u_n$$
converges, then
$$\lim_{n\to\infty} nu_n=0.$$

Prove also that
$$\sum_{n=1}^{\infty} n(u_n-u_{n+1})$$
converges.

25. Prove that the series
$$1-\frac{3}{x}+\frac{3.4}{x(x+1)}-\frac{3.4.5}{x(x+1)(x+2)}+\ldots$$
converges for real values of x only if $x>3$.

26. Examine, by means of the limit
$$\lim_{n\to\infty} n\left(\frac{a_n}{a_{n+1}}-1\right),$$
the convergence of the series

(i) $$\sum_1^{\infty} \frac{(2n)!}{2^{2n}(n+1)!\,(n-1)!},$$

(ii) $$\left\{\frac{1.3.5.\ldots.(2n-1)}{2.4.6.\ldots.(2n)}\right\}^k.$$

27. The numbers of a sequence $u_0, u_1, u_2, \ldots$ are connected by the relation

$$u_n - (k + k^{-1})\, u_{n-1} + u_{n-2} = 0 \quad (n \geqslant 2).$$

Prove that u_n is of the form

$$u_n \equiv Ak^n + Bk^{-n},$$

where A, B are constants.

Prove that, if $k > 1$ and if $u_0 = 1$, then the only value of u_1 for which u_n tends to a finite limit as n tends to infinity is $1/k$.

28. Positive numbers $x_1, x_2, x_3, \ldots$ are defined by the relations

$$x_1 = 1,$$

$$x_{n+1} = (x_n + 6)^{\frac{1}{3}} \quad (n = 1, 2, 3, \ldots).$$

Prove that, for all values of n,

$$x_n < x_{n+1} < 2.$$

Hence prove that x_n tends to a limit, and find this limit.

29. A sequence $u_1, u_2, u_3, \ldots$ is defined by the recurrence relation

$$5u_{n+1} = u_n^2 + 6.$$

Prove that $u_n \to \infty$ if $u_1 > 3$.

Prove that u_n tends to a finite limit if u_1 has any value in the range $0 \leqslant u_1 \leqslant 3$, and find the values of the limit for these values of u_1.

30. The terms S_n of a bounded sequence of real numbers satisfy the inequality

$$2s_n \leqslant s_{n-1} + s_{n+1}.$$

Prove that

$$\lim_{n \to \infty} (s_{n+1} - s_n) = 0.$$

31. Prove that, if $f(x)$ is a continuous steadily decreasing function for $x \geqslant 0$, then

$$\sum_{r=1}^{n} f(r) \leqslant \int_0^n f(x)\, dx \leqslant \sum_{r=0}^{n-1} f(r).$$

Prove that, if

$$u_n = \sum_{r=0}^{n} \frac{1}{2r+1},$$

then

$$u_n - \tfrac{1}{2} \log_e n$$

tends to a finite limit as $n \to \infty$.

32. Use the integral test to find the values of δ for which the series

$$\sum_{n=2}^{\infty} \frac{1}{n(\log n)^{1+\delta}}$$

is convergent.

33. Show that the series

$$1 - \tfrac{1}{2} + \tfrac{1}{3} - \tfrac{1}{4} + \dots,$$

$$\tfrac{1}{2} + \tfrac{1}{2}(1 - \tfrac{1}{2}) - \tfrac{1}{2}(\tfrac{1}{2} - \tfrac{1}{3}) + \tfrac{1}{2}(\tfrac{1}{3} - \tfrac{1}{4}) - \dots$$

both converge and have the same sum.

34. Show that the series

$$\frac{1}{\sqrt{8}} + \frac{1}{\sqrt{27}} + \frac{1}{\sqrt{64}} + \frac{1}{\sqrt{125}} + \dots$$

is convergent.

35. The terms of the series

$$\sum_{n=1}^{\infty} \frac{(-1)^{n-1}}{n}$$

are rearranged so that p positive terms are followed by q negative terms and so on alternately. Prove that the sum of the resulting series is

$$\log 2 + \tfrac{1}{2}\log(p/q).$$

36. Find the sum of the series

$$\sum_{n=1}^{\infty} \frac{8n-3}{n(4n-3)(4n-1)}.$$

37. Evaluate the limit, as n tends to infinity, of

$$\frac{1}{n+1} + \frac{1}{n+2} + \frac{1}{n+3} + \dots + \frac{1}{3n}.$$

38. Test for convergence the series

$$(1 + \tfrac{1}{2}) - \tfrac{2}{3} + (\tfrac{1}{4} + \tfrac{1}{5}) - \tfrac{2}{6} + (\tfrac{1}{7} + \tfrac{1}{8}) - \tfrac{2}{9} + \dots.$$

39. Show that, as $n \to \infty$,

$$1 + \frac{1}{3} + \frac{1}{5} + \dots + \frac{1}{2n-1} - \tfrac{1}{2}\log n \to \log 2 + \tfrac{1}{2}\gamma,$$

where γ is Euler's constant.

40. The series

(a) $$u_1 - u_2 + u_3 - u_4 + \dots$$

is rearranged as

(b) $$u_1 - u_2 - u_4 + u_3 - u_6 - u_8 + u_5 - u_{10} - u_{12} + \dots,$$

and the nth partial sums of the series (a) and (b) are denoted by A_n and B_n respectively. Prove that, for any positive integer m,

$$A_{2m} - B_{3m} = \sum_{r=m+1}^{2m} u_{2r}.$$

Show that, if $u_n = n^{-\frac{1}{2}}$, the series (a) is convergent but the series (b) is divergent to $-\infty$.

CHAPTER XXIV

THE DEFINITION OF FUNCTIONS BY SERIES

The need for the work which follows arises, among other things, from the fact that, though the sum of a *finite* number of terms

$$u_1(x)+u_2(x)+\dots+u_n(x)$$

of continuous functions is continuous, and can be integrated or differentiated term by term, this is not necessarily true for *infinite* series. We are seeking conditions under which these processes can then be carried out.

1. The functions defined. The primary concern of this chapter is the use of an infinite series as the definition of a function. The terms of a series

$$u_1(x)+u_2(x)+u_3(x)+\dots$$

are functions of a variable x, and the nth partial sum is $S_n(x)$, where

$$S_n(x) \equiv u_1(x)+u_2(x)+\dots+u_n(x).$$

It may happen that, for a *given* value of x_0 of x, the sum $S_n(x_0)$ tends to a limit $S(x_0)$ as n tends to infinity; the series then *converges*, for $x=x_0$, to the value $S(x_0)$. The fact of convergence at a particular value x_0, however, does not by any means ensure convergence at another value x_1; this is familiar in the case of the geometric series

$$1+x+x^2+x^3+\dots,$$

which converges when $|x|<1$ but not when $|x|\geqslant 1$.

When the series converges for a *range* of values of x, say for values in the interval

$$a<x<b,$$

its sum $S(x)$ is a function defined for the various values of x in that interval, so that

$$S(x) \equiv u_1(x)+u_2(x)+u_3(x)+\dots \quad (a<x<b).$$

We are now to examine some properties of $S(x)$.

The work of the preceding chapter may be applied to any series in which x has a given value x_0, but care must be taken once x is allowed to vary; difficulties then arise which we shall meet almost immediately.

NOTE. Though the notation x for the independent variable implies that we are thinking mainly of real variables, the ideas developed in this chapter apply equally well to complex variables, EXCEPT that *the discussion of integration and differentiation in* §§ 4, 5 *necessarily presupposes real variables only*. We do not consider in these volumes the more general theory of complex variables.

2. The problem of continuity. Some typical difficulties in the theory of functions defined by means of infinite series may be illustrated through an examination of the conditions for continuity. We give first an example to show that, *even when each of the functions $u_1(x), u_2(x), u_3(x), \ldots$ is continuous in the interval and when the sum to infinity $S(x)$ exists at each point of the interval, it is nevertheless possible for $S(x)$ to be a* DISCONTINUOUS *function of x*:

Consider the series

$$\frac{x^2}{1+x^2}+\frac{x^2}{(1+x^2)^2}+\frac{x^2}{(1+x^2)^3}+\ldots+\frac{x^2}{(1+x^2)^r}+\ldots.$$

This is a geometric progression with first term $x^2/(1+x^2)$ and ratio $1/(1+x^2)$, so that the sum of its first n terms is

$$S_n=\frac{x^2/(1+x^2)}{1-\{1/(1+x^2)\}}\left\{1-\left(\frac{1}{1+x^2}\right)^n\right\}.$$

Since $1+x^2 \neq 0$, we may multiply numerator and denominator by it; thus

$$S_n=\frac{x^2}{x^2}\left\{1-\left(\frac{1}{1+x^2}\right)^n\right\}.$$

For all values of x *other than zero* we may cancel x^2 from numerator and denominator, giving

$$S_n(x)=1-\left(\frac{1}{1+x^2}\right)^n \quad (x \neq 0).$$

Since $x^2>0$, the term $(1+x^2)^{-n}$ tends to zero, so that we have the formula

$$S(x)=1 \quad (x \neq 0).$$

On the other hand, direct substitution in the given series leads to the relation

$$S_n(0)=0,$$

so that

$$S(0)=0.$$

The function $S(x) \equiv \dfrac{x^2}{1+x^2}+\dfrac{x^2}{(1+x^2)^2}+\dfrac{x^2}{(1+x^2)^3}+\ldots$

thus has the value
$$\begin{cases} S(x)=1 & (x \neq 0), \\ S(0)=0. \end{cases}$$

It is therefore discontinuous at the origin.

3. Uniformity of convergence. The concept of uniformity, at which we have arrived, may be exhibited from several points of view; in particular, the problem of continuity raised in the preceding paragraph gives an excellent example of just how it is forced upon us. Consider, more generally, the series

$$u_1(x)+u_2(x)+u_3(x)+\ldots$$

with nth partial sum $S_n(x_0)$ at a particular value $x=x_0$, where

$$S_n(x_0) \equiv u_1(x_0)+u_2(x_0)+\ldots+u_n(x_0).$$

If $S_n(x_0)$ approaches the limit $S(x_0)$ as n tends to infinity, then, given ϵ, there exists a number N such that

$$|S(x_0)-S_n(x_0)|<\epsilon$$

whenever $n>N$.

Suppose now that the series is also convergent for a value x_1 fairly near to x_0. Then, given ϵ', there exists a number M such that

$$|S(x_1)-S_n(x_1)|<\epsilon'$$

whenever $n>M$.

We assume that $u_1(x), u_2(x), \ldots$ are continuous functions of x in a certain interval $a<x<b$ containing x_0, x_1. It is easy to prove directly from the definition of continuity that the sum of the *finite* number of terms

$$S_n(x) \equiv u_1(x)+u_2(x)+\ldots+u_n(x)$$

is also continuous. Thus, given η, there exists a number ζ such that

$$|S_n(x_1)-S_n(x_0)|<\eta$$

whenever $|x_1-x_0|<\zeta$.

(At first sight we seem all set for the argument, expressed informally:

$$\text{‘}S(x_1) \underset{\text{(convergence)}}{\doteq} S_n(x_1) \underset{\text{(continuity)}}{\doteq} S_n(x_0) \underset{\text{(convergence)}}{\doteq} S(x_0),$$

so that, regarding x_0 as given and x_1 as approaching it,

$$S(x_1) \rightarrow S(x_0);$$

hence $S(x)$ is continuous.’

But there is a flaw in the argument, as we must now explain by expanding it in detail.)

We regard x_0 as given and x_1 as varying to approach it. The inequality,

$$|S(x_0) - S_n(x_0)| < \epsilon$$

whenever $n > N$, is already established, but the next step requires care. It is true, as stated, that

$$|S(x_1) - S_n(x_1)| < \epsilon'$$

whenever $n > M$, *but that value of M depends, in general, not only on ϵ'* BUT ALSO ON x_1, and this dependence may cause trouble. For the admittedly finite number M cannot be regarded as given; it may be forced to increase as x_1 varies approaching x_0, and, in awkward cases, that increase may carry it above any preassigned number, however large.

We may follow this process for the series

$$\frac{x^2}{1+x^2} + \frac{x^2}{(1+x^2)^2} + \frac{x^2}{(1+x^2)^3} + \dots$$

already quoted. Write $x_0 \equiv 0$ and let x_1 be a number close to zero. For given ϵ', there exists indeed a number M such that

$$|S(x_1) - S_n(x_1)| < \epsilon'$$

whenever $n > M$; for, since (as proved)

$$S_n(x_1) = 1 - \left(\frac{1}{1+x_1^2}\right)^n$$

and

$$S(x_1) = 1,$$

the inequality is

$$\frac{1}{(1+x_1^2)^n} < \epsilon',$$

so that

$$(1+x_1^2)^n > 1/\epsilon'$$

and

$$n > \frac{\log(1/\epsilon')}{\log(1+x_1^2)}.$$

We therefore take M sufficiently large to ensure that

$$M > \frac{\log(1/\epsilon')}{\log(1+x_1^2)}.$$

But as x_1 approaches zero, the value of $\log(1+x_1^2)$ also approaches zero; and so the value required for M increases without limit, carrying n with it.

To enable us to avoid such problems, where M increases without bound, we introduce the concept of uniformity, with the definition:

The function $S_n(x)$ is said to CONVERGE UNIFORMLY *to $S(x)$ in a given interval $a<x<b$ if, given ϵ, a number $N(\epsilon)$ can be found* INDEPENDENT OF *x such that, whenever n is greater than N, the inequality*

$$|S(x)-S_n(x)|<\epsilon$$

holds for all x satisfying the relation $a<x<b$.

It is now easy to establish a condition for the continuity of the sum $S(x)$, which raised difficulties in the preceding paragraph:

If

$$S_n(x)\equiv u_1(x)+u_2(x)+\ldots+u_n(x),$$

where $u_1(x), u_2(x), \ldots$ are continuous functions of x, and if $S_n(x)$ converges UNIFORMLY *to the function $S(x)$ in the interval $a<x<b$, then $S(x)$ is a continuous function of x in the interval.*

To prove this, let ϵ be a given positive number. By the uniformity of convergence of $S_n(x)$ to $S(x)$, there exists a number $N(\epsilon)$, independent of x, such that

$$|S(x)-S_n(x)|<\tfrac{1}{3}\epsilon$$

whenever $n>N$ and $a<x<b$. Take, say, $n=N+1$.

Now $S_{N+1}(x)$ is the sum of a *finite* number of continuous terms, and is therefore itself continuous. Hence there exists a number ζ such that

$$|S_{N+1}(x_1)-S_{N+1}(x_0)|<\tfrac{1}{3}\epsilon$$

whenever $|x_1-x_0|<\zeta$. The uniformity inequality then gives, for the given x_0 and for such a value of x_1, the two relations

$$|S(x_0)-S_{N+1}(x_0)|<\tfrac{1}{3}\epsilon,$$

$$|S(x_1)-S_{N+1}(x_1)|<\tfrac{1}{3}\epsilon.$$

Thus

$$\begin{aligned}
&|S(x_1)-S(x_0)|\\
&\quad\equiv|\{S(x_1)-S_{N+1}(x_1)\}+\{S_{N+1}(x_1)-S_{N+1}(x_0)\}+\{S_{N+1}(x_0)-S(x_0)\}|\\
&\quad\leqslant|S(x_1)-S_{N+1}(x_1)|+|S_{N+1}(x_1)-S_{N+1}(x_0)|+|S_{N+1}(x_0)-S(x_0)|\\
&\quad<\tfrac{1}{3}\epsilon+\tfrac{1}{3}\epsilon+\tfrac{1}{3}\epsilon=\epsilon.
\end{aligned}$$

The function $S(x)$ is therefore continuous at each point of the given interval $a<x<b$.

It is important to observe that *a function $S(x)$ defined as the limit of a sequence $\{S_n(x)\}$ may be continuous even though the convergence is* NOT *uniform*. We exhibit this phenomenon by an illustration.

ILLUSTRATION 1. *To verify that the function*

$$S(x) \equiv \lim_{n\to\infty} S_n(x),$$

where

$$S_n(x) \equiv nx e^{-nx^2},$$

is continuous at the origin although $S_n(x)$ is NOT *uniformly convergent in any interval including it.*

When x is not zero,

$$\begin{aligned} S(x) &= \lim_{n\to\infty} \frac{nx}{e^{nx^2}} \\ &= \lim_{n\to\infty} \frac{nx}{1+nx^2+\frac{1}{2}n^2x^4+\ldots} \\ &= 0. \end{aligned}$$

Also we have, for all values of n, the equality

$$S_n(0) = 0,$$

so that

$$S(0) = 0.$$

Hence $S(x)$ is continuous at the origin.

On the other hand, we require, for uniformity of convergence in an interval including the origin, the condition that, given ϵ, there exists $N(\epsilon)$, independent of x, such that

$$|S(x) - S_n(x)| < \epsilon$$

whenever $n > N$. Since $S(x) = 0$, this condition is

$$|nx e^{-nx^2}| < \epsilon.$$

Now if, for any N proposed, we were to take the values, say, $n = 2N$, $x = 1/2N$ this would involve the inequality

$$(2N)(1/2N)\, e^{-1/2N} < \epsilon,$$

or

$$e^{-1/2N} < \epsilon.$$

But if $N > 1$, then $e^{-1/2N} > e^{-\frac{1}{2}}$; and so the inequality $e^{-1/2N} < \epsilon$ cannot be satisfied for small ϵ. Hence the convergence is not uniform.

NOTE. The moral of this illustration is that the condition of *uniformity* of convergence is quite a severe restriction on a sequence. A series may very well have a continuous sum without satisfying the condition of uniform convergence. Conversely, however, uniformity does imply continuity.

ILLUSTRATION 2. *The function x^n.* The way in which the concept of uniformity enters into apparently simple problems may be illustrated by means of the sequence $S_1, S_2, S_3, \ldots$ defined by the relation

$$S_n(x) \equiv x^n$$

in the interval $(0, 1)$. When $0 < x < 1$, the limit of x^n is 0, so that

$$S(x) = 0 \quad (0 < x < 1).$$

On the other hand, $S_n(1) = 1$ always, so that

$$S(x) = 1 \quad (x = 1).$$

The diagram (Fig. 148) illustrates the curves

$$y = x^n$$

for $n = 1, 2, 3, 4, \ldots$, and it is seen that, as n becomes larger and larger, the curves tend to 'settle into the corner OMA'; for very large n the curve lies close to the x-axis from O till very near to M, after which it climbs steeply towards $A(1, 1)$ as x approaches the value 1.

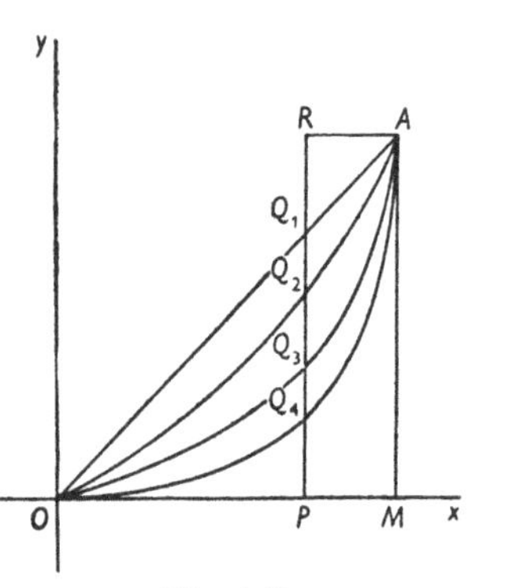

Fig. 148

Take any particular value of x, say $x = p$, close to $x = 1$. Let P be the point $(p, 0)$; suppose that the straight line $x = p$ meets the curve $y = x^n$ in the point Q_n and the straight line $y = 1$ in R. For any *given* value of n, the value of p^n is represented by PQ_n, and the 'rise' which the function has to make in order to reach the value 1 at $x = 1$ is represented by Q_nR.

Now the point about the non-uniformity of the convergence of x^n to its limit is this, that, as n increases, Q_n 'falls' towards P, leaving an ever-increasing 'rise' required to reach the level of A; and *this difficulty is not avoided by moving P closer to M.* So long as P is not actually at M, the point Q_n will move down and the 'gap' lengthen as n increases.

As n tends to infinity, x^n tends to the function $f(x)$ defined by the relations

$$\begin{cases} f(x) = 0 \quad \text{when} \quad 0 \leqslant x < 1, \\ f(1) = 1. \end{cases}$$

The uniform convergence of x^n in the interval $0 \leqslant x \leqslant 1$ would require that, given ϵ, a number $N(\epsilon)$ could be found such that

$$|x^n - f(x)| < \epsilon$$

for all x such that $0 \leqslant x \leqslant 1$ and for all $n > N$. In particular, it would require

$$x^n < \epsilon$$

for all x such that $0 < x < 1$ and for all $n > N$. This inequality requires

$$n \log x < \log \epsilon,$$

or, since $\log x$ and $\log \epsilon$ are both negative,

$$n > \frac{\log \epsilon}{\log x}.$$

But $\log x \to 0$ as $x \to 1$ and so this inequality *cannot hold* for all x in the interval $0 < x < 1$ when n exceeds any stated N whatsoever. The convergence is therefore not uniform in the interval $0 \leqslant x \leqslant 1$.

EXAMPLE

Consider similarly the sequence for which

$$S_n(x) = x^n(1 - x^n)$$

in the interval $(0, 1)$. Show that $S_n(x) \to 0$ for all values of x, including the end-points $x = 0$ and $x = 1$, but that the convergence is not uniform. Draw the graph

$$y = x^n(1 - x^n);$$

show that y has a maximum value of $\frac{1}{4}$ when $x = 1/\sqrt[n]{2}$, and indicate how this illustrates the non-uniformity.

4. Integration* of a function defined as an infinite series. Our dealings with a function $S(x)$ defined by means of the infinite series

$$S(x) \equiv u_1(x) + u_2(x) + u_3(x) + \ldots$$

will be concerned, not only with continuity, but also with differentiability and integrability. It turns out to be more convenient to begin with the latter:

To prove that, if $u_1(x), u_2(x), \ldots$ are continuous, and if the infinite series

$$u_1(x) + u_2(x) + u_3(x) + \ldots$$

* See the note on p. 109.

converges UNIFORMLY *to* $S(x)$ *in a certain (finite) interval, then*

$$\int_p^q S(x)\,dx = \int_p^q u_1(x)\,dx + \int_p^q u_2(x)\,dx + \int_p^q u_3(x)\,dx + \dots$$

for a range of integration (p, q) *lying within the given interval of uniform convergence.*

We say that the series is then *integrated term by term*; briefly,

$$\int_p^q \left\{ \sum_1^\infty u_i(x) \right\} dx = \sum_1^\infty \int_p^q u_i(x)\,dx.$$

Suppose that ϵ is a given positive number. By the definition of uniform convergence, a number N can be found, *independent of* x, such that

$$|S(x) - u_1(x) - u_2(x) - \dots - u_n(x)| < \epsilon$$

for all $n > N$. Thus we may write $S(x)$ in the form

$$S(x) = u_1(x) + \dots + u_n(x) + \eta_n(x),$$

where $|\eta_n(x)| < \epsilon$ whenever $n > N$. Hence

$$\int_p^q S(x)\,dx = \int_p^q u_1(x)\,dx + \dots + \int_p^q u_n(x)\,dx + \int_p^q \eta_n(x)\,dx,$$

where, since $|\eta_n(x)|$ is less than ϵ for all x in the interval simultaneously,

$$\left| \int_p^q \eta_n(x)\,dx \right| < \epsilon(q-p).$$

Since ϵ may be chosen arbitrarily small, and the value of n then selected to exceed the consequent value of N, however large, it follows that

$$\int_p^q S(x)\,dx = \int_p^q u_1(x)\,dx + \int_p^q u_2(x)\,dx + \dots.$$

COROLLARY. *The series*

$$\int_p^x u_1(t)\,dt + \int_p^x u_2(t)\,dt + \dots$$

converges uniformly to $\displaystyle\int_p^x S(t)\,dt.$

This is essentially the step

$$\left| \int_p^q \eta_n(x)\,dx \right| < \epsilon(q-p)$$

of the above proof.

Since
$$u_1(x)+u_2(x)+\ldots$$
converges uniformly, the sum $S(x)$ is a continuous function of x, and so (Volume I, p. 87) the function
$$F(x)\equiv\int_p^x S(t)\,dt$$
has differential coefficient $F'(x)\equiv S(x)$.

5. Differentiation* of a function defined as an infinite series. *To prove that, if $u_1'(x), u_2'(x), \ldots$ are continuous functions of x, and if the infinite series*
$$u_1(x)+u_2(x)+\ldots$$
converges to $S(x)$ in a certain interval, then
$$S'(x)=u_1'(x)+u_2'(x)+u_3'(x)+\ldots$$
whenever the series on the right converges uniformly.

The series is then said to be *differentiated term by term*; briefly,
$$\frac{d}{dx}\sum_1^\infty u_i(x)=\sum_1^\infty \frac{d}{dx}u_i(x).$$

We are given that, for a certain interval,
$$u_1'(x)+u_2'(x)+u_3'(x)+\ldots$$
converges uniformly, say to a sum $g(x)$. Thus, by the preceding theorem, since $u_1'(t), u_2'(t), \ldots$ are continuous,
$$\int_p^x g(t)\,dt=\int_p^x u_1'(t)\,dt+\int_p^x u_2'(t)\,dt+\ldots$$
$$=\{u_1(x)-u_1(p)\}+\{u_2(x)-u_2(p)\}+\ldots.$$
Now we have, by definition of the partial sums, the relation
$$\{u_1(x)-u_1(p)\}+\{u_2(x)-u_2(p)\}+\ldots+\{u_n(x)-u_n(p)\}$$
$$\equiv S_n(x)-S_n(p);$$
and we know that the limit of the difference $S_n(x)-S_n(p)$ is equal to the difference $S(x)-S(p)$ of the limits. Hence, for the infinite series,
$$\{u_1(x)-u_1(p)\}+\{u_2(x)-u_2(p)\}+\ldots=S(x)-S(p).$$

* See the note on p. 109.

Thus
$$\int_p^x g(t)\,dt = S(x) - S(p).$$

Since $g(x)$, being the sum of a uniformly convergent series of continuous functions, is continuous, it follows, by the remark at the end of the preceding paragraph, that
$$g(x) \equiv S'(x),$$
so that
$$S'(x) \equiv u_1'(x) + u_2'(x) + u_3'(x) + \ldots.$$

6. Tests for uniform convergence. It is often a matter of some difficulty to decide whether or not a given sequence or series converges uniformly. Sometimes the negative answer can be obtained directly by a contradiction of the definition itself, and this possibility should always be kept in mind. We give, more positively, a brief account of one well-known method for establishing uniformity of convergence.

The tests available for a sequence may be expected to be simpler than those for a series, since an explicit formulation for the sum of a series may not be obtainable. We therefore begin with a test for a *sequence*, remembering that it can always be used for a series in which the sum to n terms is known.

THE 'MAXIMUM VALUE' TEST FOR A SEQUENCE. Suppose that a sequence $S_1, S_2, S_3, \ldots$ is such that
$$S_n(x) \to S(x)$$
for each value of x in an interval $a < x < b$. Denote by U_n the greatest value for given n of $|S(x) - S_n(x)|$ in the interval. We prove that *the sequence is uniformly convergent to $S(x)$ if, and only if,*
$$U_n \to 0$$
as $n \to \infty$.

(i) Suppose that the convergence is uniform. Then, given ϵ, we can find $N(\epsilon)$, independent of x, such that
$$|S(x) - S_n(x)| < \epsilon,$$
for all x satisfying the relation $a < x < b$, whenever $n > N$. But U_n is a value (the greatest, in fact) of $|S(x) - S_n(x)|$ in the interval, so that, in particular,
$$U_n < \epsilon$$
for $n > N$. Thus
$$U_n \to 0.$$

(ii) Suppose that $U_n \to 0$. Then, given ϵ, we can find $N(\epsilon)$ such that

$$U_n < \epsilon$$

whenever $n > N$. But, by definition of U_n,

$$|S(x) - S_n(x)| \leqslant U_n$$

for all values of x in the interval. Hence

$$|S(x) - S_n(x)| < \epsilon$$

independently of x, so that the convergence is uniform.

ILLUSTRATION 3. *The sequence* nxe^{-nx^2}. If

$$S_n(x) \equiv nxe^{-nx^2},$$

then (p. 113)

$$S(x) \equiv 0.$$

Thus

$$|S(x) - S_n(x)| = nxe^{-nx^2}$$

for $x \geqslant 0$.

Write

$$y = nxe^{-nx^2},$$

so that

$$y' = ne^{-nx^2} - 2n^2x^2e^{-nx^2}$$
$$= ne^{-nx^2}(1 - 2nx^2).$$

Hence y has a turning value when $x = \sqrt{(1/2n)}$, and it is easy to verify that y is then a maximum. Thus, if U_n is the greatest value of $|S(x) - S_n(x)|$ for given x, then

$$U_n = n\sqrt{(1/2n)}\, e^{-\frac{1}{2}}$$
$$= e^{-\frac{1}{2}}\sqrt{(n/2)}.$$

Thus U_n does not tend to zero, and so the convergence is not uniform.

A similar argument holds when x is negative.

A useful test to decide that a *series* is uniformly convergent, is provided by WEIERSTRASS'S '*M*-TEST':

The series

$$u_1(x) + u_2(x) + u_3(x) + \dots$$

is uniformly convergent in a given interval if there exists a series of POSITIVE CONSTANTS

$$M_1 + M_2 + M_3 + \dots$$

with the properties that (i) *the series* $M_1 + M_2 + M_3 + \dots$ *is convergent*; (ii) $|u_n(x)| \leqslant M_n$ *for all values of* n *(save, possibly, a finite number at the beginning) and for all values of* x *in the interval.*

Given a positive number ϵ, we can find a number $N(\epsilon)$ such that, for all values of n greater than N,

$$M_{n+1}+M_{n+2}+M_{n+3}+\ldots<\epsilon.$$

Thus

$$\begin{aligned}|u_{n+1}(x)+u_{n+2}(x)+\ldots| &\leqslant |u_{n+1}(x)|+|u_{n+2}(x)|+\ldots \\ &\leqslant M_{n+1}+M_{n+2}+\ldots \\ &<\epsilon\end{aligned}$$

independently of x. Thus the given series is uniformly convergent.

REVISION EXAMPLES XIX

1. The function $F(x)$ is defined for real values of x by the equation

$$F(x)=\int_0^x \frac{dt}{1+t^2}.$$

Prove from the definition that $F(x)$ is a continuous, odd function, strictly increasing as x increases.

By expressing the integrand in the form

$$\frac{1}{1+t^2}=1-t^2+t^4-\ldots+(-t^2)^n+R_n(t),$$

prove that, for $-1\leqslant x\leqslant 1$,

$$F(x)=\sum_{n=0}^{\infty}(-1)^n\frac{x^{2n+1}}{2n+1}.$$

2. Show that the series

$$1-t^3+t^6-t^9+\ldots=\frac{1}{1+t^3}$$

is not uniformly convergent in the range $0\leqslant t\leqslant 1$, but that it can nevertheless be integrated term by term over this range.

Evaluate

$$\int_0^1 \frac{dt}{1+t^3},$$

and deduce that

$$1-\tfrac{1}{4}+\tfrac{1}{7}-\tfrac{1}{10}+\ldots=\{\pi+3^{\frac{1}{2}}\log 2\}/3^{\frac{3}{2}}.$$

3. Prove that the sequence

$$\frac{x^n-x^{n+1}}{1+x^{2n}}$$

is uniformly convergent when $0\leqslant x\leqslant 2$.

Is

$$\frac{x^2}{1+x^{2n}}$$

uniformly convergent in the same interval?

4. Show that $$1-x^{2n} > 2nx^n(1-x)$$

when $0 \leqslant x < 1$, and hence that the series

$$\sum_{n=1}^{\infty} \frac{x^n(1-x)}{n(1-x^{2n})}$$

is uniformly convergent in the interval $0 \leqslant x < 1$.

5. Given that the series $\sum_{r=0}^{\infty} a_r$ is convergent, prove that the series

$$\sum_{r=0}^{\infty} \frac{a_r x^r}{r!}$$

is uniformly convergent in every interval $0 \leqslant x \leqslant X$.

6. Show that the infinite series

$$\sum_{n=1}^{\infty} \frac{\sin nx}{n^p}$$

converges uniformly for all real values of x if $p > 1$.

7. Investigate the uniformity of convergence of the series

$$\sum_{n=1}^{\infty} \{e^{-n^2x^2} - e^{-(n-1)^2x^2}\}$$

for $0 \leqslant x \leqslant 1$.

8. If $$S_n(x) \equiv n^p x^n (1-x)^2,$$

determine the values of p for which, as $n \to \infty$, $S_n(x)$ tends to zero uniformly for $0 \leqslant x \leqslant 1$.

9. Discuss the uniformity of convergence in the interval $0 \leqslant x \leqslant 1$ of the series

$$\sum_{n=1}^{\infty} \frac{x^n(1-x)}{\sqrt{n}}.$$

10. If $$s_n(x) = \frac{n^p x}{(1+nx^2)^2},$$

find the range of values of p for which $s_n(x)$ tends to a limit as n tends to infinity, uniformly in $0 \leqslant x \leqslant 1$.

11. Prove that, if

$$0 \leqslant a_n < 1 \quad (n = 1, 2, 3, \ldots)$$

and
$$a_1 + a_2 + a_3 + \ldots$$

is convergent, then
$$a_1^r + a_2^r + a_3^r + \ldots$$

converges uniformly with respect to r for $r \geqslant 1$, and that its sum tends to zero as $r \to +\infty$.

Prove that, if r is an integer,

$$\lim_{r \to +\infty} \sum_{n=1}^{\infty} \sin^{2r} \frac{\pi}{2n} = 1.$$

12. State and prove sufficient conditions that

$$\int_a^b \lim_{n \to \infty} s_n(x)\, dx = \lim_{n \to \infty} \int_a^b s_n(x)\, dx.$$

If $a = 0$, $b = 1$, and $\quad s_n(x) = n^p(1-x)\, x^n,$

find for what values of p (real) (i) the conditions are satisfied, (ii) the above equation holds.

13. Show that the series

$$\sum_{n=1}^{\infty} (1-x)\{(n-1)\, x^n - n x^{n+1}\}$$

is uniformly convergent in the interval

$$0 \leqslant x \leqslant 1 - \delta \quad (0 < \delta < 1)$$

but not in the interval $0 \leqslant x \leqslant 1$.

If $S_n(x)$ is the nth partial sum of this series, show that, for every $n \geqslant 1$, a value of x such that $0 < x < 1$ can be found for which

$$|S_n(x)| \geqslant \tfrac{4}{27}.$$

14. Discuss the convergence and uniform convergence of the series

$$1 + \sum_{n=1}^{\infty} \frac{x^n(x-1)}{(1+x^n)(1+x^{n+1})}$$

in the interval $0 \leqslant x \leqslant 1$.

15. If
$$f_n(x) \equiv \frac{n}{n^2 + x^2},$$

prove that
$$\lim_{n \to \infty} f_n(x) = 0$$

uniformly for all real x, but that

$$\lim_{n\to\infty}\int_{-\infty}^{\infty} f_n(x)\,dx \neq 0.$$

If

$$g_n(x)=\frac{nx}{1+n^2x^2},$$

prove that

$$\lim_{n\to\infty}\int_0^1 g_n(x)\,dx=0,$$

and that

$$\lim_{n\to\infty} g_n(x)=0$$

for $0\leqslant x\leqslant 1$, but not uniformly.

16. Prove that the infinite series

$$\sum_{n=1}^{\infty}\frac{1}{n}\sin(\pi x^n)$$

converges uniformly in the interval $0\leqslant x\leqslant 1-\delta<1$.

17. If $S_n(x)$ is the sum to n terms of the series

$$\frac{x(x-1)}{(1)(x+1)(2x+1)}+\frac{x(2x-1)}{(x+1)(2x+1)(3x+1)}+\frac{x(3x-1)}{(2x+1)(3x+1)(4x+1)}+\ldots,$$

show that, for $x\geqslant 0$, $S_n(x)\to S(x)\equiv\dfrac{x}{2(x+1)}$

as $n\to\infty$, but that $S\left(\dfrac{1}{n}\right)-S_n\left(\dfrac{1}{n}\right)=\dfrac{1}{4}$.

What is the significance of the last result?

18. Given that $S_n(x)\equiv\dfrac{n}{n^2+(x-n)^2},$

prove that $S_n(x)$ converges to its limit $S(x)$ uniformly for all values of x, but that

$$\lim_{n\to\infty}\int_0^{\infty} S_n(x)\,dx\neq\int_0^{\infty} S(x)\,dx.$$

19. Prove that, if

$$s_n(x)=\frac{x^n}{x^{2n}+1},\quad t_n(x)=(x-1)\,s_n(x),$$

then (i) the sequence $s_n(x)$ converges uniformly for $x \geqslant 1+\delta$ for any positive δ, but not for $x \geqslant 1$; (ii) the sequence $t_n(x)$ converges uniformly for $x \geqslant 1$.

20. Show that if $\sum_{n=1}^{\infty} |u_n(x)|$ is uniformly convergent in an interval $[a, b]$, then so also is $\sum_1^{\infty} u_n(x)$.

By taking $u_n(x) = (-1)^n (1-x^2) x^n$ and $[0, 1]$ for $[a, b]$, show that it may happen that $\sum_1^{\infty} u_n(x)$ is uniformly convergent in $[a, b]$ and $\sum_1^{\infty} |u_n(x)|$ is convergent for all x in $[a, b]$, but $\sum_1^{\infty} |u_n(x)|$ not uniformly convergent in $[a, b]$.

21. Prove that $\sum_1^{\infty} \frac{x}{(nx+1)\{(n+2)x+1\}}$ is uniformly convergent in $x \geqslant \delta$ for all $\delta > 0$, but is not uniformly convergent in $x \geqslant 0$.

CHAPTER XXV

POWER SERIES

1. Introductory. A series in the form

$$a_0+a_1z+a_2z^2+a_3z^3+\ldots,$$

where the coefficients $a_0, a_1, a_2, \ldots$ are constants, is called a POWER SERIES, and, when convergent, defines a function $S(z)$. The coefficients and the variable z are, in general, complex; for real series we shall often denote the variable by the letter x.

It is possible to have power series which converge

(i) for no value of z except $z=0$;

(ii) for a limited range of values of z;

(iii) for all values of z.

Typical (real) series are:

(i) $1+(2x)^2+(4x)^4+\ldots+(2rx)^{2r}+\ldots.$

For a given non-zero value of x we have

$$u_n^{1/n}=(2nx)^2=4n^2x^2,$$

so that
$$u_n^{1/n}\to\infty.$$

Hence, by Cauchy's test (p. 85), the series is not convergent except for $x=0$.

(ii) $1+x+x^2+\ldots+x^r+\ldots,$

which, as we know, converges for all values of x in the range

$$-1<x<1,$$

but not otherwise.

(iii) $1+x^2+\dfrac{x^4}{2!}+\dfrac{x^6}{3!}+\ldots+\dfrac{x^{2r}}{r!}+\ldots.$

For a given (non-zero) value of x we have

$$\frac{u_n}{u_{n+1}}=\frac{n+1}{x^2},$$

so that
$$\frac{u_n}{u_{n+1}}\to\infty.$$

Hence, by D'Alembert's test (p. 84), the series converges for all values of x.

2. The radius of convergence. Suppose that the power series

$$a_0 + a_1 z + a_2 z^2 + \dots$$

is known to converge for a certain value Z, so that the series

$$a_0 + a_1 Z + a_2 Z^2 + \dots$$

is convergent. We prove that *the series converges absolutely for every value of z such that*

$$|z| < |Z|.$$

If the series converges, the terms are necessarily bounded; let M be a number which exceeds the greatest of the (positive) numbers

$$|a_0|, \quad |a_1 Z|, \quad |a_2 Z^2|, \quad \dots,$$

so that, for all n, $$|a_n||Z|^n < M.$$

We therefore have the inequality

$$\begin{aligned} |a_n z^n| &= |a_n||z|^n \\ &\leqslant M\left|\frac{z}{Z}\right|^n. \end{aligned}$$

Now the series $$M\left\{1 + \left|\frac{z}{Z}\right| + \left|\frac{z}{Z}\right|^2 + \dots\right\}$$

is convergent, and so, by the Comparison Test, the series

$$|a_0| + |a_1 z| + |a_2 z^2| + \dots$$

also converges, so that the given series

$$a_0 + a_1 z + a_2 z^2 + \dots$$

converges absolutely.

We can prove, by easy extension, that, *if the series is known to diverge for a certain value Z, then it diverges for every value of z such that*

$$|z| > |Z|.$$

For if the series converged for the value z, it would, by the preceding theorem, converge also for the value Z, since

$$|Z| < |z|.$$

But this contradicts the datum of divergence for the value Z.

We are now in a position to define the *radius of convergence* of the power series $$a_0 + a_1 z + a_2 z^2 + a_3 z^3 + \dots.$$

If we exclude for the moment the two cases in which the series (i) converges for no value of z except zero, (ii) converges for all values of z, then we have a series which is known to converge for certain values of z but not for others. Further, we have just established that the moduli of the values for convergence cannot exceed the moduli for non-convergence. It follows (but the proof requires deeper theory than we have at our disposal, and will only be sketched informally) that *there exists a number R such that the series converges whenever*

$$|z| < R$$

and diverges whenever

$$|z| > R.$$

The behaviour for $|z| = R$ requires particular examination for each individual series.

To indicate the method of proof, let us suppose that the series converges inside the circle $|z| = p$ and diverges outside the circle $|z| = q$. Then, necessarily,

$$p \leqslant q.$$

If $p < q$, consider a point $z = \zeta$ on the circle $|z| = \frac{1}{2}(p+q)$. If the series converges for ζ, it converges for $|z| < \frac{1}{2}(p+q)$; if the series diverges for ζ, it diverges for $|z| > \frac{1}{2}(p+q)$. The 'gap' $q-p$ between the radii for convergence and divergence has, in either case, been cut by half. Proceeding in this way, we can steadily diminish the 'gap', halving it at each step. It thus shrinks to zero, so that the regions for convergence and divergence are separated by the circle which we have called $|z| = R$.

The number R is called the RADIUS OF CONVERGENCE of the given series. The word 'radius' is, of course, drawn from the representation of the complex variable z in an Argand diagram. The series converges for all values of z within the circle $|z| = R$ and diverges for all values outside, with the possibility of either convergence or divergence on the circumference. The circle $|z| = R$ is called the CIRCLE OF CONVERGENCE of the series.

It is convenient to say that a series which diverges everywhere, except for $z = 0$, has zero radius of convergence, and that a series which converges everywhere has infinite radius.

Analogous definitions may be given for a series

$$a_0 + a_1(z-z_0) + a_2(z-z_0)^2 + a_3(z-z_0)^3 + \ldots,$$

centred, as it were, upon the fixed point z_0 rather than the origin. The substitution

$$z - z_0 = \zeta$$

brings it to the earlier type.

3. Two formulae for the radius of convergence. The convergence tests of D'Alembert and Cauchy for series of positive terms serve to derive two formulae for the evaluation of the radius of convergence R of the series

$$a_0+a_1z+a_2z^2+a_3z^3+\dots.$$

The series of moduli is

$$|a_0|+|a_1||z|+|a_2||z|^2+|a_3||z|^3+\dots.$$

(i) USE OF D'ALEMBERT'S TEST. Since

$$\frac{u_n}{u_{n+1}}=\frac{|a_n||z|^n}{|a_{n+1}||z|^{n+1}}$$

$$=\left|\frac{a_n}{a_{n+1}}\right|\left|\frac{1}{z}\right|,$$

the series of moduli converges if

$$\lim_{n\to\infty}\left|\frac{a_n}{a_{n+1}}\right|\left|\frac{1}{z}\right|>1,$$

or
$$|z|<\lim_{n\to\infty}\left|\frac{a_n}{a_{n+1}}\right|,$$

provided that the limit exists.

The given series converges for all values of z for which this inequality holds, and so, by definition, the radius of convergence is given by the formula

$$R=\lim_{n\to\infty}\left|\frac{a_n}{a_{n+1}}\right|,$$

provided that the limit exists.

If this limit is zero, the series converges for $z=0$ only; if it is 'infinite', the series converges for all values of z.

(ii) USE OF CAUCHY'S TEST. Since

$$u_n^{1/n}=|a_n|^{1/n}|z|,$$

the series of moduli converges if

$$\lim_{n\to\infty}|a_n|^{1/n}|z|<1,$$

or
$$|z|<\lim_{n\to\infty}|a_n|^{-1/n},$$

provided that the limit exists. Hence, as for D'Alembert's test,

$$R = \lim_{n\to\infty} |a_n|^{-1/n},$$

provided that the limit exists.

If this limit is zero, the series converges for $z=0$ only; if it is 'infinite', the series converges for all values of z.

4. Uniformity of convergence.

Suppose that the series

$$a_0 + a_1 z + a_2 z^2 + \dots$$

has radius of convergence R, so that it defines a function

$$S(z) \equiv a_0 + a_1 z + a_2 z^2 + \dots$$

for all values of z for which $|z| < R$. The series may or may not converge for $|z| = R$.

Let us, to avoid uncertainty at the value R itself, fix a positive number r of value definitely less than R though possibly close to it. Thus

$$r < R,$$

and $S(z)$ is defined whenever

$$|z| \leqslant r.$$

We prove first that *the power series converges* UNIFORMLY *for all values of z such that* $|z| \leqslant r$. This is, in fact, an immediate consequence* of Weierstrass's M-test (p. 119); for the modulus $|a_n z^n|$ of the term $a_n z^n$ is less than or equal to the corresponding term $|a_n| r^n$ of the convergent series

$$|a_0| + |a_1| r + |a_2| r^2 + \dots,$$

and so the uniformity of convergence is established.

COROLLARY. *The function $S(z)$ is a continuous function of z when* $|z| \leqslant r$.

This is merely a particular case of the more general result proved on p. 112.

5. Integration and differentiation.

We restrict the work of this paragraph to series in which the variable z is real and therefore denoted by the letter x. The theory on which we shall lean (pp. 115–8)

* It is easy to confirm that the proof given for this theorem holds whether z is real or complex.

was established for real variable only. Similar results do, indeed, hold for complex z, but they involve the ideas of differentiation and integration with respect to a complex variable, which are not considered in this book.

(i) THE INTEGRATION OF THE (REAL) POWER SERIES

$$S(x) \equiv a_0 + a_1 x + a_2 x^2 + \dots.$$

If R is the radius of convergence, and r any number such that $r < R$, then (p. 129) the series is uniformly convergent for $|x| \leqslant r$ and therefore (p. 115) integrable term by term. Hence

$$\int_0^x S(t)\,dt = a_0 x + \tfrac{1}{2} a_1 x^2 + \tfrac{1}{3} a_2 x^3 + \dots \quad (|x| < R).$$

(ii) THE DIFFERENTIATION OF THE (REAL) POWER SERIES

$$S(x) \equiv a_0 + a_1 x + a_2 x^2 + \dots.$$

In order to show that the power series may be differentiated term by term, it is necessary (p. 117) to establish the uniform convergence of the differentiated series. By Weierstrass's M-test, this would follow from the convergence of the series

$$|a_1| + 2|a_2| r + 3|a_3| r^2 + \dots.$$

Since $r < R$, we can find a number s such that

$$r < s < R.$$

The inequality $s < R$ gives the convergence of the series

$$|a_0| + |a_1| s + |a_2| s^2 + \dots.$$

Hence the terms of this series are bounded, and so there exists a number k such that

$$|a_n| s^n < k$$

for all values of n. Hence

$$n|a_n| r^{n-1} < \frac{nk}{r}\left(\frac{r}{s}\right)^n,$$

so that the terms of the series

$$|a_1| + 2|a_2| r + 3|a_3| r^2 + \dots$$

are less than the corresponding terms of the series

$$\frac{k}{r}\left\{\left(\frac{r}{s}\right) + 2\left(\frac{r}{s}\right)^2 + 3\left(\frac{r}{s}\right)^3 + \dots\right\}.$$

But this series converges by D'Alembert's test (p. 84), since

$$\lim_{n\to\infty} \frac{u_n}{u_{n+1}} = \lim_{n\to\infty} \frac{n}{n+1}\left(\frac{s}{r}\right) = \frac{s}{r}$$

$$> 1 \quad (s > r).$$

The series $$a_1 + 2a_2x + 3a_3x^2 + \ldots$$
thus converges uniformly to $S'(x)$ for $|x| \leqslant r$, so long as $r < R$. Hence *the series*

$$a_1 + 2a_2x + 3a_3x^2 + \ldots$$

converges uniformly to $S'(x)$, with the radius of convergence R of the given series.

This result may be applied inductively any number of times to obtain the successive differential coefficients $S''(x), S'''(x), \ldots$ by term-by-term differentiations of the corresponding series.

6. The sum of two series. Let

$$S(z) \equiv a_0 + a_1z + a_2z^2 + \ldots,$$

$$T(z) \equiv b_0 + b_1z + b_2z^2 + \ldots$$

be two given series with radii of convergence A, B respectively. Then each series converges for values of z such that

$$|z| < r,$$

provided that $r < A$, $r < B$. Thus (p. 80) *the two series can be added term by term for values of z within the* SMALLER *of the two circles of convergence*, so that, in that region,

$$S(z) + T(z) = (a_0 + b_0) + (a_1 + b_1)z + (a_2 + b_2)z^2 + \ldots.$$

7. 'Equating coefficients.' Suppose that a function $S(x)$ is represented by two distinct power series

$$S(x) \equiv a_0 + a_1x + a_2x^2 + \ldots,$$

$$S(x) \equiv b_0 + b_1x + b_2x^2 + \ldots$$

with (non-zero) radii of convergence A, B. *To prove that corresponding coefficients are equal, so that*

$$a_0 = b_0, \quad a_1 = b_1, \quad a_2 = b_2, \quad \ldots.$$

By the result proved in § 6 we have, for $|x|<A$, $|x|<B$, the relation

$$\begin{aligned}(a_0-b_0)+(a_1-b_1)x+(a_2-b_2)x^2+\dots\\ \equiv S(x)+\{-S(x)\}\\ \equiv 0;\end{aligned}$$

say

$$c_0+c_1x+c_2x^2+\dots\equiv 0.$$

Differentiating term by term, and so restricting this proof to real values of x (compare p. 129), we obtain successively the relations, all true for $|x|<A$, $|x|<B$,

$$\begin{aligned}c_1+\quad 2c_2x+\quad 3c_3x^2+\dots&\equiv 0,\\ 2c_2+\quad 3.2c_3x+\quad 4.3c_4x^2+\dots&\equiv 0,\\ 3.2c_3+4.3.2c_4x+5.4.3c_5x^2+\dots&\equiv 0,\end{aligned}$$

and so on. Putting $x=0$ in these identities, we obtain the relations

$$c_0=0,\quad c_1=0,\quad c_2=0,\quad c_3=0,\quad \dots,$$

or

$$a_0=b_0,\quad a_1=b_1,\quad a_2=b_2,\quad a_3=b_3,\quad \dots.$$

8. The multiplication of power series. Let

$$a_0+a_1z+a_2z^2+\dots,$$
$$b_0+b_1z+b_2z^2+\dots$$

be two power series each convergent within the circle $|z|=\rho$, so that the radius of convergence for each is not less then ρ. We prove* that, *if A, B are the sums of the series for a given value of z ($|z|<\rho$), then the series*

$$c_0+c_1z+c_2z^2+\dots,$$

where

$$c_r\equiv a_0b_r+a_1b_{r-1}+\dots+a_rb_0,$$

also converges, to a sum $C=AB$.

Write

$$|a_n|=\alpha_n,\quad |b_n|=\beta_n,\quad |z|=r.$$

Since $r<\rho$, each given series converges absolutely, so that the series

$$P\equiv\alpha_0+\alpha_1r+\alpha_2r^2+\dots,$$
$$Q\equiv\beta_0+\beta_1r+\beta_2r^2+\dots$$

both converge.

Consider now the product P_nQ_n of the partial sums of P, Q, where

$$P_nQ_n=(\alpha_0+\alpha_1r+\dots+\alpha_nr^n)(\beta_0+\beta_1r+\dots+\beta_nr^n).$$

* The final step of the proof will be seen to have the limitation that z is real; but the result is true generally.

Then
$$P_n Q_n \leqslant PQ.$$

Write
$$\gamma_r \equiv \alpha_0\beta_r + \alpha_1\beta_{r-1} + \ldots + \alpha_r\beta_0;$$

then
$$P_n Q_n = \gamma_0 + \gamma_1 r + \ldots + \gamma_n r^n + (\text{terms in } r^{n+1}, r^{n+2}, \ldots, r^{2n}).$$

If we put
$$R_n \equiv \gamma_0 + \gamma_1 r + \ldots + \gamma_n r^n,$$

then (see also the note at the end of this section)
$$R_n \leqslant P_n Q_n,$$

so that, since $P_n Q_n \leqslant PQ$, we have
$$R_n \leqslant PQ.$$

Since, then, $\{R_n\}$ is a bounded increasing sequence of positive terms, the limit
$$R \equiv \lim_{n\to\infty} R_n$$

exists. Hence *the series*
$$\begin{aligned} C &\equiv c_0 + c_1 z + c_2 z^2 + \ldots \\ &\equiv a_0 b_0 + (a_0 b_1 + a_1 b_0) z + (a_0 b_2 + a_1 b_1 + a_2 b_0) z^2 + \ldots, \end{aligned}$$

converges absolutely, since the moduli of its terms are equal to or less than the terms of *the series* $\gamma_0 + \gamma_1 r + \gamma_2 r^2 + \ldots$.

Moreover, since the series of *positive* terms
$$\alpha_0\beta_0 + (\alpha_0\beta_1 + \alpha_1\beta_0) r + (\alpha_0\beta_2 + \alpha_1\beta_1 + \alpha_2\beta_0) r^2 + \ldots$$

converges (to R), the sum of *positive* terms
$$\alpha_0\beta_0 + \alpha_0\beta_1 r + \alpha_1\beta_0 r + \alpha_0\beta_2 r^2 + \alpha_1\beta_1 r^2 + \alpha_2\beta_0 r^2 + \ldots$$

with brackets removed also converges, so that the series
$$a_0 b_0 + a_0 b_1 z + a_1 b_0 z + a_0 b_2 z^2 + a_1 b_1 z^2 + a_2 b_0 z^2 + \ldots$$

converges absolutely. But we proved (p. 92) FOR A SERIES OF REAL TERMS that, if it converges absolutely, then the terms can be rearranged while not affecting the sum. Hence, for REAL z,
$$\begin{aligned} C &\equiv c_0 + c_1 z + c_2 z^2 + c_3 z^3 + \ldots \\ &\equiv a_0 b_0 + (a_0 b_1 + a_1 b_0) z + (a_0 b_2 + a_1 b_1 + a_2 b_0) z^2 + \ldots \\ &\equiv a_0 b_0 + (a_0 b_1 z + a_1 b_0 z + a_1 b_1 z^2) \\ &\qquad + (a_0 b_2 z^2 + a_2 b_0 z^2 + a_1 b_2 z^3 + a_2 b_1 z^3 + a_2 b_2 z^4) + \ldots, \end{aligned}$$

where terms bracketed are added to make up the successive products $a_0 b_0$, $(a_0 + a_1 z)(b_0 + b_1 z)$, $(a_0 + a_1 z + a_2 z^2)(b_0 + b_1 z + b_2 z^2)$, ...

Hence

$$C \equiv \lim_{n\to\infty} (\text{sum of first } n+1 \text{ terms as arranged})$$

$$\equiv \lim_{n\to\infty} (a_0 + a_1 z + \ldots + a_n z^n)(b_0 + b_1 z + \ldots + b_n z^n)$$

$$\equiv AB.$$

The theorem as stated is true for complex terms, but the final step, as given here, has depended on an earlier theorem whose proof involved real terms only. The general case of that theorem is, however, easily established from the consideration that a complex series converges if, and only if, its real and imaginary parts converge. The present result then follows.

NOTE. The relationship between R_n, $P_n Q_n$ may be exhibited diagrammatically. We choose in illustration the case $n = 3$. The sixteen terms in the product $P_3 Q_3$ may be written in the form of a square:

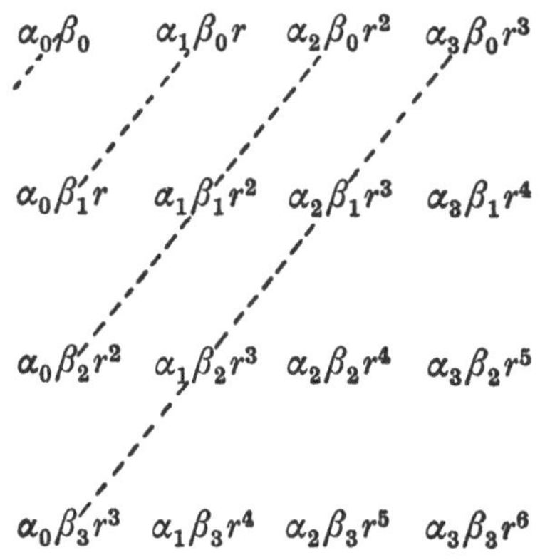

The terms γ_0, $\gamma_1 r$, $\gamma_2 r^2$, $\gamma_3 r^3$ are precisely the members of the four dotted 'diagonals'. Hence

$$R_n \equiv \text{sum of terms on dotted lines}$$

$$\leqslant \text{sum of all the terms}$$

$$\leqslant P_n Q_n.$$

REVISION EXAMPLES XX

Find the radii of convergence (possibly zero or 'infinite') of the following series:

1. $1+z+\frac{z^2}{(2!)^2}+\frac{z^3}{(3!)^2}+\frac{z^4}{(4!)^2}+\dots$

2. $1-\frac{1}{2}z^2+\frac{1}{4}z^4-\frac{1}{8}z^6+\frac{1}{16}z^8-\dots$

3. $z^2+2z^4+3z^6+4z^8+\dots$

4. $2+3z+4z^2+5z^3+\dots$

5. $z+\left(\frac{z}{2}\right)^2+2!\left(\frac{z}{3}\right)^3+3!\left(\frac{z}{4}\right)^4+\dots+(n-1)!\left(\frac{z}{n}\right)^n+\dots$

6. $z+3z^3+5z^5+7z^7+\dots$

7. $2+(1+2)z+(1+2^2)z^2+\dots+(1+2^{r-1})z^{r-1}+\dots$

8. $1+z+\left(\frac{z}{2}\right)^2+\left(\frac{z}{3}\right)^3+\left(\frac{z}{4}\right)^4+\dots$

9. $z+\frac{z^2}{2}+\frac{z^3}{3}+\frac{z^4}{4}+\dots$

10. $z+\frac{z^2}{2^2}+\frac{z^3}{3^2}+\frac{z^4}{4^2}+\dots$

11. $z+2^2z^2+3^3z^3+4^4z^4+\dots$

12. $1+\frac{(2z)^2}{2!}+\frac{(3z)^3}{3!}+\frac{(4z)^4}{4!}+\dots$

13. Prove that the series

$$1+z+z^2+z^3+\dots$$

does not converge for any value of z such that $|z|=1$.

14. Prove that the two series

$$\sum_1^\infty 3^n z^n, \quad \sum_1^\infty (3-z)^n$$

cannot both be convergent for the same value of z.

Find the values of z for which neither series is convergent.

15. State whether the condition

$$u_n \to 0 \quad \text{as} \quad n \to \infty$$

is (*a*) necessary, (*b*) sufficient for the convergence of the series

$$\text{(i)}\ \sum_1^\infty u_n, \quad \text{(ii)}\ \sum_1^\infty u_n x^n \quad (-1 < x < 1).$$

Prove the truth of each of your four statements.

16. Prove that

$$1 + x\cos\theta + x^2\cos 2\theta + \ldots + x^{n-1}\cos(n-1)\theta$$
$$= \frac{1 - x\cos\theta}{1 - 2x\cos\theta + x^2} + x^n \frac{x\cos(n-1)\theta - \cos n\theta}{1 - 2x\cos\theta + x^2},$$

and examine the convergence of the series as $n \to \infty$.

17. Prove that, if $-1 < x < 1$, then

$$\lim_{n\to\infty} nx^n = 0.$$

Prove that

$$\sum_{r=1}^{n} (a + rb)x^r = \frac{x(1-x^n)(a+b-ax) - nbx^{n-1}(1-x)}{(1-x)^2}.$$

Obtain the values of x for which the infinite series

$$(a+b)x + (a+2b)x^2 + (a+3b)x^3 + \ldots$$

converges, and find its sum.

18. Prove that, if $\Sigma a_n x^n$ converges when $x = \alpha$, it converges when $|x| < |\alpha|$.

Determine the real values of x for which the series

$$\sum_1^\infty n^2(x-1)^n,$$

$$x + x^4 + x^9 + x^{16} + \ldots$$

converge.

19. Determine the real values of x for which the series

$$\sum_{n=1}^{\infty} \frac{1.3.\ldots.(2n+1)}{2.4.\ldots.2n}(1-x^2)^n$$

converges.

20. Find for what real values of x the series

$$\sum_{n=1}^{\infty} \frac{(1-x)(2-x)(3-x)\ldots(n-x)}{2^n n!}$$

converges.

21. Prove that, if $\sum_{n=1}^{\infty} u_n$ is a series such that

$$n\left(\left|\frac{u_n}{u_{n+1}}\right|-1\right) > k > 1$$

for all values of n sufficiently large, then the series is absolutely convergent.

Consider the convergence of the series

$$\frac{1-a}{2}+\frac{(1-a)(1-2a)}{2.3}x+\frac{(1-a)(1-2a)(1-3a)}{2.3.4.}x^2+\dots$$

when a is positive.

22. Prove that, if the real power series

$$a_0+a_1x+a_2x^2+a_3x^3+\dots$$

is convergent when $x = k$, where k is positive, then it is uniformly convergent when

$$k > l \geqslant x \geqslant -l > -k.$$

Prove in this case that, if $f(x)$ is its sum, then

$$\int_0^x f(t)\,dt = a_0x+\tfrac{1}{2}a_1x^2+\tfrac{1}{3}a_2x^3+\dots.$$

(i) Find the sum of the series

$$\sum_{n=1}^{\infty} \frac{x^{2n+3}}{n(n+1)(2n+3)}$$

when $x^2 < 1$.

(ii) Prove that

$$\sum_{n=1}^{\infty} \frac{(-1)^{n-1}}{(2n-1)(2n+1)3^n} = \sqrt{3}\int_0^{1/\sqrt{3}} x\tan^{-1}x\,dx = \frac{2\pi\sqrt{3}-9}{18}.$$

CHAPTER XXVI

THE DEFINITION OF FUNCTIONS BY INTEGRALS

The subject of this chapter is the definition of a function by means of an integral. We have already (Volume II, p. 1) met an elementary example in the logarithm

$$F(x) \equiv \int_1^x \frac{dt}{t}.$$

It will be noticed that, in the early stages, we lean heavily on intuition, quoting without proof results which require a much deeper study of analysis than we have at our disposal. It is hoped that, on the one hand, the existence of the gaps in the argument has been made clear, and that, on the other, the underlying principles have been made sufficiently intelligible for the reader to accept the omissions as reasonable.

We begin with a preliminary theorem which is important in its own right.

1. The first mean-value theorem for integrals. Let $f(x)$ be a function of x, continuous in an interval $a \leqslant x \leqslant b$. Suppose, too, that $u(x)$ is a function, also continuous, which is POSITIVE throughout the interval. To prove that *there exists a number* ξ, *where* $a < \xi < b$, *such that*

$$\int_a^b f(x)\,u(x)\,dx = f(\xi)\int_a^b u(x)\,dx.$$

Since $f(x)$ is continuous in the interval (a, b), it can be proved (but we regard it as obvious intuitively, for example, from a graph) that $f(x)$ is bounded; so that numbers m, M exist such that

$$m \leqslant f(x) \leqslant M.$$

Divide the interval into n parts, not necessarily equal, at points where x assumes in turn (and increasingly) the values

$$a \equiv x_0,\ x_1,\ x_2,\ \ldots,\ x_{n-1},\quad x_n \equiv b.$$

Since $u(x)$ is positive everywhere,

$$m \sum_1^{n-1} u(x_i)\,(x_{i+1}-x_i) \leqslant \sum_1^{n-1} f(x_i)\,u(x_i)\,(x_{i+1}-x_i) \leqslant M \sum_1^{n-1} u(x_i)\,(x_{i+1}-x_i),$$

and, in the limit,

$$m\int_a^b u(x)\,dx \leqslant \int_a^b f(x)\,u(x)\,dx \leqslant M\int_a^b u(x)\,dx.$$

Hence there exists a number k between m and M such that

$$\int_a^b f(x)\,u(x)\,dx = k\int_a^b u(x)\,dx.$$

Now it can be proved, and we shall again regard it as sufficiently clear from graphical considerations, that a continuous function assumes in a given interval *every* value between its least and its greatest. If, then, we take m, M to be actually the least and greatest values of $f(x)$ in (a, b), there will exist a number ξ between a and b such that

$$f(\xi) = k.$$

Hence
$$\int_a^b f(x)\,u(x)\,dx = f(\xi)\int_a^b u(x)\,dx.$$

Corollary. By setting $u(x) \equiv 1$, we have the relation

$$\int_a^b f(x)\,dx = (b-a)\,f(\xi)$$

for some number ξ such that $a < \xi < b$.

(The serious student will require to fill in the steps which we have treated intuitively, and a text-book on analysis should be consulted.)

2. Functions defined by finite integrals with fixed limits.

If

$$f(x, t)$$

is a given function of two variables, x, t, then the integral

$$\int_a^b f(x, t)\,dt,$$

where a, b are constant, defines a certain function of x, say

$$F(x) \equiv \int_a^b f(x, t)\,dt,$$

for some range of values $\alpha \leqslant x \leqslant \beta$.

The properties with which we are most concerned are those of differentiability (with implications of continuity) and integrability. We consider them in turn.

(i) DIFFERENTIABILITY OF $F(x)$. Writing

$$F(x+h)=\int_a^b f(x+h,t)\,dt,$$

we have the relation

$$F(x+h)-F(x)=\int_a^b \{f(x+h,t)-f(x,t)\}\,dt.$$

Suppose now that the function $f(x,t)$ is such that the partial differential coefficient $\frac{\partial}{\partial x}f(x,t)$ exists.* By the mean-value theorem of the differential calculus, a number θ can be found, where $0<\theta<1$, such that

$$f(x+h,t)-f(x,t)=h\frac{\partial}{\partial x}f(x+\theta h,t),$$

so that

$$\frac{F(x+h)-F(x)}{h}=\int_a^b \frac{\partial}{\partial x}f(x+\theta h,t)\,dt.$$

Unfortunately, the number θ is not a pure constant, but its value depends on that of the parameter t. Hence difficulties about uniformity arise as we integrate from a to b on the right-hand side and let h tend to zero. We propose to omit the deeper analysis required for a detailed study, but to regard as plausible what is in fact true, that, letting h tend to zero in the last equation, *the differential coefficient of the function*

$$F(x)\equiv\int_a^b f(x,t)\,dt$$

is, in normal cases, given by the formula

$$F'(x)\equiv\int_a^b \frac{\partial}{\partial x}f(x,t)\,dt.$$

This result will be used whenever required, and quoted freely.

(ii) INTEGRABILITY OF $F(x)$. The function $f(x,t)$ being given, as before, as a continuous function of x, t throughout the region of values to be considered, we have seen (Volume III, pp. 109–15) how to define the double integral

$$\iint f(x,t)\,dx\,dt$$

* In this chapter, we shall confine our attention to the simplest cases only, and assume the existence (and, where necessary, the continuity) of all functions and their differential coefficients.

and also how to express it in the alternative forms of repeated integrals

$$\int dx \int f(x,t)\,dt, \quad \int dt \int f(x,t)\,dx$$

between suitable limits. In particular, if t ranges from a to b, and if the integration of $F(x)$ is from p to q, then the region of values is the rectangle $a \leqslant t \leqslant b$, $p \leqslant x \leqslant q$, and we have the formulae

$$\int_p^q dx \int_a^b f(x,t)\,dt = \int_a^b dt \int_p^q f(x,t)\,dx$$

$$= \iint_{\text{(rectangle)}} f(x,t)\,dx\,dt.$$

Thus, *if $F(x)$ is the function defined by the integral*

$$F(x) = \int_a^b f(x,t)\,dt,$$

then

$$\int_p^q F(x)\,dx = \int_a^b dt \int_p^q f(x,t)\,dx$$

$$= \iint_{\text{(rectangle)}} f(x,t)\,dx\,dt.$$

3. Functions defined by finite integrals with variable limits. Consider next the function

$$F(x) \equiv \int_a^u f(x,t)\,dt,$$

where a is constant and $u \equiv u(x)$ is a function of x. We shall not require a formula for integration, but the formula for differentiation is important.

We have

$$F(x+h) = \int_a^{u(x+h)} f(x+h,t)\,dt$$

$$= \int_a^{u(x)} f(x+h,t)\,dt + \int_{u(x)}^{u(x+h)} f(x+h,t)\,dt,$$

so that

$$F(x+h) - F(x) = \int_a^{u(x)} \{f(x+h,t) - f(x,t)\}\,dt + \int_{u(x)}^{u(x+h)} f(x+h,t)\,dt.$$

The first of these integrals can be dealt with exactly as indicated in § 2 (i), so we proceed directly to the second. The first mean-value theorem for integrals (Corollary, p. 139), gives the relation

$$\int_{u(x)}^{u(x+h)} f(x+h,t)\,dt = \{u(x+h)-u(x)\}\,f(x+h,\xi),$$

where ξ lies between $u(x)$ and $u(x+h)$. Hence, in all,

$$\frac{F(x+h)-F(x)}{h} = \int_a^{u(x)} \frac{f(x+h,t)-f(x,t)}{h}\,dt + \frac{u(x+h)-u(x)}{h}\,f(x+h,\xi),$$

and, as h tends to zero, we obtain the *formula of differentiation*

$$F'(x) = \int_a^{u(x)} \frac{\partial f(x,t)}{\partial x}\,dt + u'(x)\,f(x,u).$$

Corollary. Similar reasoning may be used to prove that, *if $F(x)$ is defined by the relation*

$$F(x) = \int_{u(x)}^{v(x)} f(x,t)\,dt,$$

then the differential coefficient $F'(x)$ is given by the formula

$$F'(x) = \int_{u(x)}^{v(x)} \frac{\partial f(x,t)}{\partial x}\,dt + v'(x)\,f(x,v) - u'(x)\,f(x,u).$$

Illustration 1. *To find the differential coefficient with respect to x of the function*

$$F(x) \equiv \int_{x^2}^{x^3} (1+3x^2t^2)\,dt.$$

By the above formula

$$\begin{aligned} F'(x) &= \int_{x^2}^{x^3} 6xt^2\,dt + 3x^2(1+3x^2.x^6) - 2x(1+3x^2.x^4) \\ &= 2x(x^9-x^6) + 3x^2(1+3x^8) - 2x(1+3x^6) \\ &= 11x^{10} - 8x^7 + 3x^2 - 2x. \end{aligned}$$

In this simple case, we can check by direct calculation; for

$$\begin{aligned} F(x) &= \Big[\,t\,\Big]_{x^2}^{x^3} + x^2\Big[\,t^3\,\Big]_{x^2}^{x^3} \\ &= x^3 - x^2 + x^{11} - x^8, \end{aligned}$$

so that $$F'(x) = 3x^2 - 2x + 11x^{10} - 8x^7.$$

4. Functions defined by infinite integrals. The real work of this chapter is directed to integrals, now to be studied, in which the limits may be infinite. There is close analogy with the work for infinite series, and steps will often be set out in a similar way.

Let $f(x,t)$ be a continuous function of x, t, and suppose that, when x has a given value x_0, the value of the integral $\int_a^T f(x_0, t)\,dt$ is denoted by $S_T(x_0)$, so that

$$S_T(x_0) = \int_a^T f(x_0, t)\,dt,$$

where a, T are constants. It may happen that $S_T(x_0)$ tends to a limit $S(x_0)$ as T tends to infinity; we say

$$S(x_0) \equiv \lim_{T\to\infty} S_T(x_0)$$

$$\equiv \int_a^\infty f(x_0, t)\,dt.$$

The existence of $S(x_0)$ requires that, if ϵ is any given positive number, however small, then there exists a number $N(\epsilon)$ such that

$$|\,S(x_0) - S_T(x_0)\,| < \epsilon$$

whenever T is greater than N.

THE 'BOUNDING' TEST *for integrals.* A test very similar to that given earlier (p. 82) for series may be established for integrals. Suppose that the function $f(x,t)$ is always POSITIVE. Then the integral

$$S_T(x_0) \equiv \int_a^T f(x_0, t)\,dt$$

is an increasing function of T, so that, if $S_T(x_0)$ is bounded for all values of T, the sequence $\{S_T(x_0)\}$ converges as $T\to\infty$. Hence, *if $f(x,t)$ is positive, and if the integral*

$$\int_a^T f(x_0, t)\,dt$$

is bounded for all values of T, then the integral

$$\int_a^\infty f(x_0, t)\,dt$$

exists.

5. Uniformity of convergence for infinite integrals. Suppose that the integral

$$S_T(x) \equiv \int_a^T f(x,t)\,dt$$

converges for each value of x in a given interval. We say that *the integral* CONVERGES UNIFORMLY *to a value $S(x)$ for all values of x in the interval if, given ϵ, there exists a number $N(\epsilon)$* INDEPENDENT OF *x such that*

$$|S(x) - S_T(x)| < \epsilon$$

whenever $T > N$.

It is sometimes convenient to write this inequality in the form

$$\left|\int_T^\infty f(x,t)\,dt\right| < \epsilon.$$

To prove that, *if $f(x,t)$ is a function, continuous in the region of values considered, and if $S_T(x)$ converges* UNIFORMLY *to $S(x)$ in a given interval, then $S(x)$ is a continuous function of x in the interval.*

Consider the two values x_0, x_0+h lying in the given interval, and the corresponding integrals

$$\int_a^\infty f(x_0+h,t)\,dt \equiv \int_a^T f(x_0+h,t)\,dt + \int_T^\infty f(x_0+h,t)\,dt$$

and
$$\int_a^\infty f(x_0,t)\,dt \equiv \int_a^T f(x_0,t)\,dt + \int_T^\infty f(x_0,t)\,dt.$$

Since the integral converges uniformly, we can, given a positive number ϵ, take T sufficiently large to ensure that

$$\left|\int_T^\infty f(x_0+h,t)\,dt\right| < \tfrac{1}{3}\epsilon,$$

$$\left|\int_T^\infty f(x_0,t)\,dt\right| < \tfrac{1}{3}\epsilon.$$

Keeping this value of T (which, by uniformity, does not depend on h), we can now choose h sufficiently small to ensure, by the continuity of $f(x,t)$, that

$$|f(x_0+h,t) - f(x_0,t)| < \frac{\epsilon}{3(T-a)},$$

so that
$$\left|\int_a^T \{f(x_0+h,t) - f(x_0,t)\}\,dt\right| < \tfrac{1}{3}\epsilon.$$

Hence

$$\begin{aligned}
&| S(x_0+h)-S(x_0) | \\
&= \left| \int_a^T \{f(x_0+h,t)-f(x_0,t)\}\,dt + \int_T^\infty f(x_0+h,t)\,dt - \int_T^\infty f(x_0,t)\,dt \right| \\
&\leqslant \left| \int_a^T f(x_0+h,t)-f(x_0,t)\,dt \right| + \left| \int_T^\infty f(x_0+h,t)\,dt \right| + \left| \int_T^\infty f(x_0,t)\,dt \right| \\
&\leqslant \tfrac{1}{3}\epsilon + \tfrac{1}{3}\epsilon + \tfrac{1}{3}\epsilon \\
&\leqslant \epsilon.
\end{aligned}$$

The function is therefore continuous.

6. The (finite) integration of a function defined as an infinite integral. To prove that, *if $f(x,t)$ is continuous for values considered, and if the integral*

$$\int_a^\infty f(x,t)\,dt$$

converges UNIFORMLY *to $S(x)$ in a certain interval, then*

$$\int_p^q S(x)\,dx = \int_a^\infty dt \left\{ \int_p^q f(x,t)\,dx \right\},$$

that is,
$$\int_p^q dx \left\{ \int_a^\infty f(x,t)\,dt \right\} = \int_a^\infty dt \left\{ \int_p^q f(x,t)\,dx \right\},$$

for a range of integration (p,q) lying within the given interval.

Suppose that ϵ is a given positive number. By the definition of uniform convergence, a number N can be found, *independent of x*, such that

$$\left| \int_T^\infty f(x,t)\,dt \right| < \epsilon$$

for all $T > N$. Thus

$$S(x) = \int_a^T f(x,t)\,dt + \eta_T(x),$$

where $| \eta_T(x) | < \epsilon$ whenever $T > N$. Hence

$$\int_p^q S(x)\,dx = \int_p^q dx \left\{ \int_a^T f(x,t)\,dt \right\} + \int_p^q \eta_T(x)\,dx,$$

where
$$\left| \int_p^q \eta_T(x)\,dx \right| < \epsilon(q-p).$$

But, for *constant* p, q and a, T, we have (p. 140)

$$\int_p^q dx\left\{\int_a^T f(x,t)\,dt\right\} = \int_a^T dt\left\{\int_p^q f(x,t)\,dx\right\}.$$

Hence $$\int_p^q S(x)\,dx = \int_a^T dt\left\{\int_p^q f(x,t)\,dx\right\} + \int_p^q \eta_T(x)\,dx.$$

Since ϵ may be chosen arbitrarily small, and the value of T then selected to exceed the consequent value of N, however large, it follows that

$$\int_p^q S(x)\,dx = \int_a^\infty dt\left\{\int_p^q f(x,t)\,dx\right\}.$$

COROLLARY. *The integral*

$$\int_a^\infty dt\left\{\int_p^x f(\theta,t)\,d\theta\right\}$$

is uniformly convergent (within the interval of uniform convergence of the given integral, assumed finite), and defines a function

$$\int_p^x S(\theta)\,d\theta$$

whose differential coefficient is $S(x)$.

7. Differentiation under the sign of integration. To prove that, *if* $f(x,t)$ *is continuous for values considered, and if the infinite integral*

$$\int_a^\infty f(x,t)\,dt$$

converges to $S(x)$ *in a certain interval, then*

$$S'(x) = \int_a^\infty \frac{\partial f}{\partial x}\,dt$$

whenever the integral on the right converges uniformly.

It is given that

$$\int_a^\infty \frac{\partial f}{\partial x}\,dt$$

converges uniformly, say to a value $g(x)$.

By the theorem of § 6, with $\partial f/\partial x$ in place of $f(x,t)$, we have the relation

$$\int_p^x g(\theta)\,d\theta = \int_a^\infty dt\int_p^x \frac{\partial}{\partial\theta} f(\theta,t)\,d\theta$$

for any value of x in the interval (p, q) of convergence. Thus

$$\int_p^x g(\theta)\,d\theta = \int_a^\infty \{f(x,t) - f(p,t)\}\,dt$$
$$= S(x) - S(p).$$

Hence (remembering that $g(x)$ must be continuous)

$$S'(x) = g(x)$$
$$= \int_a^\infty \frac{\partial f}{\partial x}\,dt.$$

8. A test for uniform convergence. The test which follows is closely analogous to Weierstrass's M-test given earlier (p. 119) for series:

If a function $g(t)$ can be found such that

$$\int_a^\infty g(t)\,dt$$

exists and that $$|f(x,t)| \leqslant g(t)$$

for all x (within a certain interval) whenever $t \geqslant a$, then the integral

$$\int_a^\infty f(x,t)\,dt$$

is uniformly convergent.

For we can, given ϵ, find a number T_0 *independent of x*, such that, for $T > T_0$,

$$\int_T^\infty g(t)\,dt < \epsilon,$$

so that
$$\left|\int_T^\infty f(x,t)\,dt\right| \leqslant \int_T^\infty |f(x,t)|\,dt$$
$$\leqslant \int_T^\infty g(t)\,dt$$
$$< \epsilon.$$

Hence the integral converges uniformly.

Note as a COROLLARY which is sometimes useful that, *if a positive constant M and a number $k > 1$ can be found such that*

$$|f(x,t)| < Mt^{-k}$$

for $t \geqslant a$, then the integral $\int_a^\infty f(x,t)\,dt$

is uniformly convergent.

ILLUSTRATION 2. *To prove that*

$$\int_0^\infty \frac{\sin t}{t}\,dt = \tfrac{1}{2}\pi.$$

This well-known integral illustrates several features of the theory, and the steps are given in detail. We begin with two Lemmas:

(i) *The integral*
$$\int_0^\infty \frac{e^{-xt}\sin t}{t}\,dt$$
is uniformly convergent for $x \geqslant 0$.

Denote by M a large positive integer, and consider the integral

$$R_T \equiv \int_T^{M\pi} \frac{e^{-xt}\sin t}{t}\,dt.$$

If N is the least positive integer exceeding T/π, then

$$R_T = \int_T^{N\pi} \frac{e^{-xt}\sin t}{t}\,dt + \int_{N\pi}^{M\pi} \frac{e^{-xt}\sin t}{t}\,dt$$

$$= \int_T^{N\pi} \frac{e^{-xt}\sin t}{t}\,dt + \sum_{r=N}^{M-1} \int_{r\pi}^{(r+1)\pi} \frac{e^{-xt}\sin t}{t}\,dt.$$

(See the Note below for a point of detail.)

Write
$$t = r\pi + u;$$
then
$$\int_{r\pi}^{(r+1)\pi} \frac{e^{-xt}\sin t}{t}\,dt = \int_0^\pi \frac{e^{-x(r\pi+u)}\sin(r\pi+u)}{r\pi+u}\,du$$
$$= (-1)^r \int_0^\pi \frac{e^{-xu}\sin u}{e^{xr\pi}(r\pi+u)}\,du.$$

Since $1/\{e^{xr\pi}(r\pi+u)\}$ decreases as r increases (x being positive), the integrals under the sign of summation $\sum_{r=N}^{M-1}$ have the properties:

(*a*) they alternate in sign;
(*b*) they decrease steadily in magnitude for $x \geqslant 0$;
(*c*) they tend to zero as r tends to infinity for $x \geqslant 0$.

If we denote the terms in the summation $\sum_{N}^{M-1}$ by the notation

$$u_N - u_{N+1} + u_{N+2} - u_{N+3} + \dots,$$

then, by these three conditions,

$$|u_N - u_{N+1} + u_{N+2} - \dots| < |u_N|,$$

so that $$|R_T| < \left|\int_T^{N\pi} \frac{e^{-xt}\sin t}{t}\,dt\right| + \left|\int_{N\pi}^{(N+1)\pi} \frac{e^{-xt}\sin t}{t}\,dt\right|.$$

For each integral, since $x \geqslant 0$, $T > 0$, $t > T$, we have

$$1/t < 1/T, \quad e^{-xt} \leqslant 1.$$

Hence $$|R_T| \leqslant \frac{1}{T}\int_T^{N\pi} |\sin t|\,dt + \frac{1}{T}\int_{N\pi}^{(N+1)\pi} |\sin t|\,dt,$$

and so we obtain, as an outside estimate, the inequality, independent of x,

$$|R_T| < 2\pi/T \to 0.$$

The integral therefore converges uniformly. We denote its sum by $S(x)$, so that

$$S(x) \equiv \int_0^\infty \frac{e^{-xt}\sin t}{t}\,dt.$$

NOTE. The working

$$\int_{N\pi}^{M\pi} \frac{e^{-xt}\sin t}{t}\,dt \equiv \sum_{r=N}^{M-1} \int_{r\pi}^{(r+1)\pi} \frac{e^{-xt}\sin t}{t}\,dt$$

carries the implication that the upper limit of the given integral is an exact multiple of π. But no essential limitation is involved; for the integral

$$\int_{r\pi}^{(r+1)\pi} \frac{e^{-xt}\sin t}{t}\,dt,$$

in the form $$\pm\int_0^\pi \frac{e^{-xu}\sin u}{e^{xr\pi}(r\pi+u)}\,du \quad (t = r\pi + u),$$

tends to zero as r tends to infinity. The integrand

$$\frac{e^{-xu}\sin u}{e^{xr\pi}(r\pi+u)}$$

is positive throughout the interval, so that the numerical value of the integral is decreased if the upper limit π is replaced by any smaller positive number. Thus the integral still tends to zero, and the main theorem remains undisturbed.

(ii) *The integral*

$$\int_0^\infty e^{-xt}\sin t\,dt$$

is uniformly convergent for $x \geqslant \delta > 0$, where δ is any positive number.

Consider the remainder

$$R'_T \equiv \int_T^\infty e^{-xt}\sin t\,dt.$$

We have
$$\begin{aligned}|R_T'| &\leqslant \int_T^\infty e^{-xt}|\sin t|\,dt\\ &\leqslant \int_T^\infty e^{-xt}\,dt\\ &\leqslant \frac{e^{-xT}}{x}\\ &\leqslant \frac{e^{-\delta T}}{\delta} \to 0.\end{aligned}$$

Since this inequality is independent of x, the integral converges uniformly.

We first apply these Lemmas to the interval $x \geqslant \delta > 0$ in which both integrals are uniformly convergent. The conditions (p. 146) for differentiation under the sign of integration are all satisfied, and so, differentiating the relation
$$S(x) \equiv \int_0^\infty \frac{e^{-xt}\sin t}{t}\,dt,$$
we obtain the equality
$$\begin{aligned}S'(x) &= -\int_0^\infty e^{-xt}\sin t\,dt\\ &= -\frac{1}{1+x^2}\end{aligned}$$
after an easy calculation.

Hence
$$S(x) = C - \tan^{-1} x$$
for $x \geqslant \delta > 0$. Also we have, for large values of x, the inequality
$$|S(x)| \leqslant \int_0^\infty e^{-xt}\,dt$$
since $|\sin t| < t$; so that
$$|S(x)| \leqslant \frac{1}{x}.$$

It follows that
$$\lim_{x\to\infty} S(x) = 0,$$
or
$$C - \tfrac{1}{2}\pi = 0,$$
so that
$$C = \tfrac{1}{2}\pi.$$

We therefore have the formula
$$S(x) \equiv \int_0^\infty \frac{e^{-xt}\sin t}{t}\,dt = \tfrac{1}{2}\pi - \tan^{-1} x,$$

established for $x \geqslant \delta > 0$.

Finally, we have proved that the integral $S(x)$ itself (though not the integral giving its differential coefficient) is uniformly convergent for $x \geqslant 0$, so that $S(x)$ is continuous at $x=0$; moreover, $\tan^{-1} x$ is also continuous at $x=0$, tending to the value zero. Hence

$$S(0) = \tfrac{1}{2}\pi;$$

that is,

$$\int_0^\infty \frac{\sin t}{t}\, dt = \tfrac{1}{2}\pi.$$

9. The gamma (factorial) and beta functions. The GAMMA FUNCTION $\Gamma(x+1)$, known alternatively as the FACTORIAL FUNCTION $x!$, is defined by the integral

$$\Gamma(x+1) \equiv x! \equiv \int_0^\infty t^x e^{-t}\, dt.$$

(Both names are in use; the gamma notation is probably still the more usual.)

(i) *Convergence.* We must establish the convergence of the integral both at zero and at infinity. To do this, we split it into the two parts

$$\int_0^1 t^x e^{-t}\, dt, \quad \int_1^\infty t^x e^{-t}\, dt,$$

and examine each in turn.

We begin with

$$\int_\delta^1 t^x e^{-t}\, dt,$$

with $0 < \delta < 1$.

Now the integral is *not* convergent for $x \leqslant -1$. For when t lies in the interval $(\delta, 1)$, it is subject to the inequality

$$e^{-t} \equiv \frac{1}{e^t} > \frac{1}{e},$$

so that

$$\begin{aligned}\int_\delta^1 t^x e^{-t}\, dt &> \int_\delta^1 t^x e^{-1}\, dt \\ &= e^{-1}\left[\frac{t^{x+1}}{x+1}\right]_\delta^1 (x \neq -1) \\ &= \frac{1}{e(x+1)}\{1-\delta^{x+1}\}.\end{aligned}$$

If $x+1 < 0$, say $x+1 = -\eta$, where η is positive, then the relation is

$$\int_\delta^1 t^x e^{-t}\, dt > \frac{1}{e\eta}\left\{\frac{1}{\delta^\eta} - 1\right\};$$

and the right-hand side becomes indefinitely large as $\delta \to 0$, so that the integral does not exist for $x < -1$. The case $x = -1$ is similar, but simpler.

If, however, $x > -1$, we use the inequality

$$e^{-t} \equiv \frac{1}{e^t} < 1 \quad (t > 0),$$

giving
$$\int_\delta^1 t^x e^{-1} dt < \int_\delta^1 t^x dt = \left[\frac{t^{x+1}}{x+1}\right]_\delta^1$$

$$= \frac{1}{(x+1)}\{1 - \delta^{x+1}\},$$

where the exponent $x+1$ is now positive. Hence

$$\int_\delta^1 t^x e^{-t} dt < \frac{1}{x+1}.$$

The integral therefore remains bounded as $\delta \to 0$; moreover, the integrand $t^x e^{-t}$ is always positive for $t > 0$. Thus (p. 143) the gamma function is convergent at the lower (zero) limit.

Consider next
$$\int_1^T t^x e^{-t} dt,$$

with $T > 1$.

Denote by k the (fixed) integer which is the first to be greater than both 1 and x. Then, when $t > 1$,

$$t^x < t^k$$

and
$$\frac{1}{e^t} = \frac{1}{1 + t + \ldots + \{t^{k+2}/(k+2)!\} + \ldots}$$

$$< \frac{1}{\{t^{k+2}/(k+2)!\}} = \frac{(k+2)!}{t^{k+2}}.$$

Hence
$$t^x e^{-t} < \frac{(k+2)!}{t^2},$$

so that
$$\int_1^T t^x e^{-t} dt < (k+2)! \int_1^T \frac{dt}{t^2}$$

$$< (k+2)!\left\{1 - \frac{1}{T}\right\}$$

$$< (k+2)!.$$

The integral therefore remains *bounded* as $T \to \infty$; also the integrand $t^x e^{-t}$ is positive in the interval. Thus (p. 143) the gamma function is convergent at the upper (infinite) limit.

Hence *the function*

$$\Gamma(x+1) \equiv x! \equiv \int_0^\infty t^x e^{-t}\, dt$$

has a value for each value of x, provided that $x > -1$.

(ii) *The 'factorial' property.* The basic property, to which the name 'factorial function' is due, is expressed by the relation

$$\left.\begin{array}{l} \Gamma(x+2) = (x+1)\,\Gamma(x+1), \\ \text{or (what is equivalent)} \\ (x+1)! = (x+1)\,x!. \end{array}\right\} \quad (x > -1).$$

For

$$\begin{aligned} (x+1)! &= \int_0^\infty t^{x+1} e^{-t}\, dt \\ &= \Big[-t^{x+1}e^{-t}\Big]_0^\infty + \int_0^\infty (x+1)\, t^x e^{-t}\, dt \\ &= 0 + (x+1)\int_0^\infty t^x e^{-t}\, dt \\ &= (x+1)\, x!. \end{aligned}$$

Note also that

$$\begin{aligned} \Gamma(1) \equiv 0! &\equiv \int_0^\infty e^{-t}\, dt \\ &= 1. \end{aligned}$$

Hence *if x is a positive integer,*

$$\Gamma(x+1) \equiv x! \equiv x(x-1)(x-2)\ldots 3.2.1.$$

The BETA FUNCTION $B(x, y)$ is defined by the integral

$$B(x, y) = \int_0^1 t^{x-1}(1-t)^{y-1}\, dt.$$

By reasoning similar to that just given for the factorial function, we may prove that convergence for $t=0$ requires $x > 0$ and that convergence for $t=1$ requires $y > 0$. Thus we assume that $x > 0$, $y > 0$.

Writing $t \equiv 1 - t'$ and then suppressing dashes, we have the relation

$$B(x, y) = \int_0^1 (1-t)^{x-1}\, t^{y-1}\, dt,$$

so that

$$B(x, y) = B(y, x).$$

Writing $t = \sin^2\theta$, we also have the relation

$$B(x, y) = 2\int_0^{\frac{1}{2}\pi} \sin^{2x-1}\theta \cos^{2y-1}\theta\, d\theta.$$

We prove now that *the beta and factorial functions are connected by the relation*

$$B(x,y)=\frac{\Gamma(x)\,\Gamma(y)}{\Gamma(x+y)}\qquad (x,y>0),$$

or (in factorial notation)

$$B(x+1,y+1)=\frac{x!\,y!}{(x+y+1)!}\qquad (x+1,y+1>0).$$

Consider the product

$$x!\,y!=\int_0^\infty t^x e^{-t}\,dt\int_0^\infty u^y e^{-u}\,du$$

$$=\int_0^\infty\int_0^\infty t^x u^y e^{-(t+u)}\,dt\,du.$$

Write $t=\xi^2$, $u=\eta^2$. Then

$$x!\,y!=4\int_0^\infty\int_0^\infty \xi^{2x+1}\eta^{2y+1}e^{-(\xi^2+\eta^2)}\,d\xi\,d\eta.$$

In terms of polar coordinates ($\xi=r\cos\theta$, $\eta=r\sin\theta$) in the (ξ,η) plane, we have, subject to convergence (see below),

$$x!\,y!=4\int_0^\infty\int_0^{\frac{1}{2}\pi} r^{2x+2y+3}e^{-r^2}\cos^{2x+1}\theta\,\sin^{2y+1}\theta\,dr\,d\theta$$

$$=4\int_0^\infty r^{2x+2y+3}e^{-r^2}\,dr\int_0^{\frac{1}{2}\pi}\cos^{2x+1}\theta\,\sin^{2y+1}\theta\,d\theta.$$

The first integral, on writing $r^2\equiv v$, is

$$\frac{1}{2}\int_0^\infty v^{x+y+1}e^{-v}\,dv\equiv\tfrac{1}{2}(x+y+1)!\qquad (x+y+1>0),$$

and the second, by the above, is

$$\tfrac{1}{2}B(x+1,y+1).$$

Hence
$$x!\,y!=(x+y+1)!\,B(x+1,y+1),$$

or
$$B(x+1,y+1)=\frac{x!\,y!}{(x+y+1)!}.$$

To justify the transformation from (ξ,η) to the polar form (r,θ), observe that the double integral is the limit of the integral

$$\int_0^N\int_0^N \xi^{2x+1}\eta^{2y+1}e^{-(\xi^2+\eta^2)}\,d\xi\,d\eta$$

over the square of vertices $(0,0)$, $(N,0)$, (N,N), $(0,N)$, whereas the integral

$$\int_0^N \int_0^{\frac{1}{2}\pi} r^{2x+2y+3} e^{-r^2} \cos^{2x+1}\theta \sin^{2y+1}\theta \, dr \, d\theta$$

is over the positive quadrant of the circle $r=N$. The limiting values will be the same provided that

$$\lim_{N\to\infty} \iint \xi^{2x+1}\eta^{2y+1} e^{-(\xi^2+\eta^2)} \, d\xi \, d\eta = 0$$

over the area of the square not covered by the circle.

Now the value of the integrand throughout this area is less than $(2N)^{2x+1}(2N)^{2y+1}/e^{N^2}$, and the area itself is equal to $N^2 - \frac{1}{4}\pi N^2$. Hence the value of the integral is less than

$$\frac{(2N)^{2x+2y+2}(1-\frac{1}{4}\pi)N^2}{e^{N^2}},$$

or

$$2^{2x+2y+2}(1-\tfrac{1}{4}\pi)N^{2x+2y+4}/e^{N^2},$$

which tends to zero as N tends to infinity.

The transformation is therefore justified.

COROLLARY. If $x = -\frac{1}{2}$, $y = -\frac{1}{2}$, then

$$\begin{aligned}\frac{(-\frac{1}{2})!\,(-\frac{1}{2})!}{0!} &= B(\tfrac{1}{2},\tfrac{1}{2}) \\ &= 2\int_0^{\frac{1}{2}\pi} \sin^0\theta \cos^0\theta \, d\theta \\ &= 2\int_0^{\frac{1}{2}\pi} d\theta \\ &= \pi.\end{aligned}$$

Hence (being positive) $\Gamma(\frac{1}{2}) \equiv (-\frac{1}{2})! = \sqrt{\pi}$.

ILLUSTRATION 3. *Dirichlet integrals.* The use of the gamma functions enables us to evaluate a class of integrals of which we give a particular example:

To evaluate the integral

$$\iiint f(x+y+z)\, x^{p-1} y^{q-1} z^{r-1} \, dx\, dy\, dz \quad (p,q,r>0)$$

over the volume of the tetrahedron bounded by the planes $x=0$, $y=0$, $z=0$, $x+y+z=1$.

Write

$$x+y+z=u, \quad y+z=uv, \quad z=uvw,$$

so that $$x=u(1-v),\quad y=uv(1-w),\quad z=uvw$$

and $$u=x+y+z,\quad v=\frac{y+z}{x+y+z},\quad w=\frac{z}{y+z}.$$

The Jacobian of the transformation is

$$\frac{\partial(x,y,z)}{\partial(u,v,w)}\equiv\begin{vmatrix} 1-v & -u & 0 \\ v(1-w) & u(1-w) & -uv \\ vw & wu & uv \end{vmatrix}=u^2v.$$

Moreover u, v, w are all positive and, for varying values of x, y, z within the tetrahedron, take independently all sets of values between 0 and 1; hence the new volume of integration is the cube $0\leqslant u\leqslant 1$, $0\leqslant v\leqslant 1$, $0\leqslant w\leqslant 1$. The integral is thus

$$\iiint f(u)\{u(1-v)\}^{p-1}\{uv(1-w)\}^{q-1}\{uvw\}^{r-1}u^2v\,du\,dv\,dw,$$

$$=\int_0^1 f(u)\,u^{p+q+r-1}\,du\int_0^1 v^{q+r-1}(1-v)^{p-1}\,dv\int_0^1 w^{r-1}(1-w)^{q-1}\,dw,$$

$$=\int_0^1 f(u)\,u^{p+q+r-1}\,du\,B(q+r,p)\,B(r,q)$$

$$=\frac{\Gamma(q+r)\,\Gamma(p)}{\Gamma(q+r+p)}\cdot\frac{\Gamma(r)\,\Gamma(q)}{\Gamma(r+q)}\int_0^1 f(u)\,u^{p+q+r-1}\,du$$

$$=\frac{\Gamma(p)\,\Gamma(q)\,\Gamma(r)}{\Gamma(p+q+r)}\int_0^1 f(u)\,u^{p+q+r-1}\,du.$$

REVISION EXAMPLES XXI

1. Prove, by differentiation with respect to the parameter α, that
$$\int_0^\pi \frac{\log(1+\cos\alpha\cos\theta)}{\cos\theta}\,d\theta=\pi(\tfrac{1}{2}\pi-\alpha).$$

2. Show, by differentiation with respect to α, that, if $\alpha\geqslant\alpha_0>0$, then
$$\int_0^\infty \frac{e^{-\alpha x}\sin x}{x}\,dx=K-\tan^{-1}\alpha,$$

where K is a constant.

Find the value of K by considering the value of the integral when $\alpha\to\infty$.

3. Prove that $$\int_0^\infty e^{-yx}\sin x\,dx = \frac{1}{1+y^2},$$

and evaluate $$\int_0^\infty \frac{e^{-yx}\sin x}{x}\,dx,$$

stating in each case the values of y for which the result holds. Justify your statements.

Deduce that $$\int_0^\infty \frac{\sin x}{x}\,dx = \tfrac{1}{2}\pi.$$

With the aid of the last result, verify (and justify your method) that, if
$$\phi(t) \equiv \int_0^\infty \frac{\sin(xt)}{x(a^2+x^2)}\,dx,$$

then $$\phi''(t) - a^2\phi(t) = \text{constant}.$$

Deduce that, if $t > 0$,
$$\phi(t) = \frac{\pi}{2a^2}(1-e^{-at}).$$

4. Prove that, if $a > 0$, $x \geqslant 0$,
$$\int_0^\infty \frac{\cos ux}{a^2+u^2}\,du = \frac{\pi}{2a}e^{-ax}.$$

5. Evaluate the integral
$$\int_0^\infty \frac{dx}{(1+a^2x^2)(1+b^2x^2)}$$

and show that it is uniformly convergent for all $a \geqslant 0$ if b is fixed and positive.

Deduce by integration the values of
$$\int_0^\infty \frac{\tan^{-1}x\,dx}{x(1+b^2x^2)}, \quad \int_0^\infty \left(\frac{\tan^{-1}x}{x}\right)^2 dx.$$

6. Prove that, if $$u = \int_0^\infty e^{-x^2}\cos 2xy\,dx,$$

then $$\frac{du}{dy} = -2yu,$$

and, assuming that $u = \frac{1}{2}\sqrt{\pi}$ when $y = 0$, deduce that
$$u = \tfrac{1}{2}\sqrt{\pi}\,e^{-y^2}.$$

7. Apply the formula

$$\frac{d}{d\alpha}\int_a^b f(x,\alpha)\,dx=\int_a^b \frac{\partial}{\partial\alpha}f(x,\alpha)\,dx$$

to show that

$$\int_0^{\frac{1}{2}\pi}\cos x\tan^{-1}(\sinh\alpha\,\sin x)\,dx=\tan^{-1}(\sinh\alpha)-\operatorname{cosech}\alpha\,\log\cosh\alpha.$$

8. Prove that $\int_0^{\pi}\log(1-2\alpha\cos x+\alpha^2)\,dx$

has the same value for all values of α in $-1<\alpha<1$.

9. Prove that, if $-1<\alpha<1$,

$$\int_0^{\frac{1}{2}\pi}\operatorname{cosec}\theta\log\left\{\frac{1+\alpha\sin\theta}{1-\alpha\sin\theta}\right\}d\theta=\pi\sin^{-1}\alpha,$$

the principal value of the inverse sine being taken.

10. Show that the integral

$$I\equiv\int_0^{\infty}\frac{x\sin xy}{x^2+a^2}\,dx$$

is uniformly convergent in any range $0<\alpha\leqslant y\leqslant\beta$.

If

$$J\equiv\int_0^{\infty}\frac{\cos xy}{x^2+a^2}\,dx,$$

prove that, when $a>0$ and $y>0$,

$$I=-\frac{\partial J}{\partial y}=-\frac{a}{y}\left(\frac{\partial J}{\partial a}+\frac{J}{a}\right).$$

11. Prove that, if $A>0$, $A+B>0$,

$$\int_0^{\frac{1}{2}\pi}\frac{dx}{(A+B\cos^2x)^{n+1}}=\frac{(-1)^n\pi}{2.n!}\frac{\partial^n}{\partial A^n}\left\{\frac{1}{\sqrt{(A^2+AB)}}\right\}.$$

Deduce that, when a and b are positive,

$$\int_0^{\frac{1}{2}\pi}\frac{dx}{(a\sin^2x+b\cos^2x)^2}=\frac{\pi(a+b)}{4(ab)^{\frac{3}{2}}}.$$

12. If $$\phi(x)\equiv\int_0^{\infty}(1-e^{-x^2y^2})\frac{dy}{y^2},$$

show that $\phi(x)/x$ is constant for $x>0$.

Calculate $\phi'(x)$ and hence evaluate

$$\int_0^\infty (1-e^{-y^2})\frac{dy}{y^2}.$$

13. Prove that, if the integral

$$\int_0^\alpha \frac{f(x)\,dx}{(\alpha-x)^{\frac{1}{2}}}$$

is independent of α, then $x^{\frac{1}{2}}f(x)$ is a constant.

14. Evaluate the integral

$$\int_0^\pi \frac{\log(1-\alpha^2\sin^2 x)}{\sin x}\,dx,$$

where $-1<\alpha<1$.

15. Prove that, if α and β are numerically less than $\frac{1}{2}\pi$,

$$\int_0^\pi \frac{\log(1+\sin\alpha\cos x)}{1+\sin\beta\cos x}\,dx = 2\pi\sec\beta\log\{\sec\tfrac{1}{2}\beta\cos\tfrac{1}{2}(\alpha+\beta)\}.$$

16. Prove that, if $-1<\alpha<1$,

$$\int_0^{\frac{1}{2}\pi} \log(1+\cos x\cos\alpha\pi)\sec x\,dx = \tfrac{1}{8}\pi^2(1-4\alpha^2).$$

17. Prove by means of the formula (under appropriate conditions)

$$\int_a^b dy\int_0^\infty \phi(x,y)\,dx = \int_0^\infty dx\int_a^b \phi(x,y)\,dy,$$

that (i) $\displaystyle\int_0^\infty \frac{e^{-x}\sin bx}{x}\,dx = \tan^{-1} b,$

(ii) $\displaystyle\int_0^\infty \frac{e^{-x}\sinh bx}{x}\,dx = \tfrac{1}{2}\log\left(\frac{1+b}{1-b}\right)$ if $|b|<1$.

18. If

$$u(y) \equiv \int_0^\infty e^{-x^2}\sin 2xy\,dx,$$

prove that

$$\frac{du}{dy}+2yu=1,$$

and deduce that

$$u(y)=e^{-y^2}\int_0^y e^{x^2}\,dx.$$

19. Prove that, if $\displaystyle f(x)\equiv\int_0^1 \frac{\cos xt}{\sqrt{(1-t^2)}}\,dt,$

then

$$f''(x)+x^{-1}f'(x)+f(x)=0.$$

20. Prove that, if $f(x)$ is a decreasing function of x, then

$$\frac{1}{b-a}\int_a^b f(t)\,dt$$

is decreased when a or b or both are increased.

21. Change the order of integration in the integral

$$\int_0^1 \left\{\int_0^{\sqrt{(1-x^2)}} \frac{dy}{(e^y+1)\sqrt{(1-x^2-y^2)}}\right\} dx,$$

and hence evaluate the integral.

22. Prove that, if

(i) $f(x,y)$ is a continuous function of (x,y) for $x \geqslant 0$, $y \geqslant 0$,

(ii) $\lim\limits_{a\to\infty}\int_0^a dx \int_a^\infty f(x,y)\,dy = \lim\limits_{a\to\infty}\int_0^a dy\int_a^\infty f(x,y)\,dx,$

(iii) $\int_0^\infty dx\int_0^\infty f(x,y)\,dy$ exists,

then $$\int_0^\infty dy\int_0^\infty f(x,y)\,dx = \int_0^\infty dx\int_0^\infty f(x,y)\,dy.$$

Prove that the three conditions are satisfied by

$$f(x,y) = \begin{cases} e^{-x}(x-y)^{\frac{1}{2}}y^{\frac{1}{2}} & (0<y<x), \\ 0 & \text{(otherwise)}, \end{cases}$$

and, by dealing with both sides of the deduced relation, prove that

$$\int_0^\infty e^{-x}x^{\frac{1}{2}}\,dx = \tfrac{1}{2}\sqrt{\pi}.$$

23. Prove that $$\int_0^1 \frac{dx}{(1-x^{\frac{1}{3}})^{\frac{1}{3}}} = \frac{81}{40}.$$

24. Prove that $$\int_0^1 \frac{x(x-x^2)^{\frac{1}{2}}}{(1+x)^4}\,dx = \tfrac{1}{128}\pi\sqrt{2}.$$

25. Prove that $$\int_0^1 \frac{dx}{\sqrt{(1-x^4)}} = \frac{1}{8}\sqrt{\left(\frac{2}{\pi}\right)}\,\{\Gamma(\tfrac{1}{4})\}^2.$$

26. Show that $$B(x,x) = 2^{1-2x}B(x,\tfrac{1}{2}),$$

and deduce that $\pi^{\frac{1}{2}}\Gamma(2x) = 2^{2x-1}\Gamma(x)\,\Gamma(x+\tfrac{1}{2}).$

Prove also that

$$\Gamma(\tfrac{1}{2})\,\Gamma(2x) = 2^{2x-1}\Gamma(x)\,\Gamma(x+\tfrac{1}{2}).$$

27. D is the area in the xy plane defined by the inequalities $x>0$, $y>0$, $x^{2m}+y^{2n}<1$, where m, n are positive but not necessarily integers. Prove that

$$\iint_D xy\,dx\,dy = \frac{1}{4(m+n)}\,\frac{\Gamma(m^{-1})\,\Gamma(n^{-1})}{\Gamma(m^{-1}+n^{-1})}.$$

(Write $x=u^{1/m}$, $y=v^{1/n}$.)

28. If $p>0$, $q>0$, $b>a>0$, evaluate

$$\int_0^a x^{p-1}(a-x)^{q-1}\,(b-x)^{-p-q}\,dx.$$

29. By substituting $x+y=u$, $y=uv$, or otherwise, prove that the value of

$$\iint \{xy(1-x-y)\}^{\frac{1}{2}}\,dx\,dy,$$

taken over the area of the triangle bounded by the lines $x=0$, $y=0$, $x+y=1$, is $2\pi/105$.

CHAPTER XXVII

THE SOLUTION OF DIFFERENTIAL EQUATIONS IN TERMS OF INFINITE SERIES

It is familiar that the solution of the differential equation

$$\frac{d^2y}{dx^2}+y=0$$

can be written in the form

$$y=A\left(1-\frac{x^2}{2!}+\frac{x^4}{4!}-\dots\right)+B\left(x-\frac{x^3}{3!}+\frac{x^5}{5!}-\dots\right),$$

where A and B are arbitrary constants. In this simple case, the solution has been expressed in terms of two infinite series, which represent the known functions $\cos x$ and $\sin x$ respectively; but often—and this is the purpose of the chapter—equations which cannot be solved in finite terms (at least, by functions already studied) may be solved instead by means of ascertainable infinite series, whose properties can then be examined as required.

There are thus three problems:

(i) How can such infinite series be determined?

(ii) When determined, under what conditions (for example, of convergence) are they really solutions?

(iii) What properties do they possess?

We confine ourselves mainly to the first of these tasks, the finding of such series in actual practice, and we restrict attention to the linear differential equation of the second order

$$\frac{d^2y}{dx^2}+P(x)\frac{dy}{dx}+Q(x)\,y=0.$$

It is found convenient to extend the scope of inquiry at once, and to search for more general solutions of the form

$$\begin{aligned}y&\equiv a_0x^c+a_1x^{c+1}+a_2x^{c+2}+\dots\\&\equiv\sum_0^\infty a_\lambda x^{c+\lambda},\end{aligned}$$

where the constant exponent c is to be determined, and where the first coefficient a_0 is not zero.

The method consists in differentiating the proposed series formally, substituting formally in the given equation, and then equating to zero the coefficients of the various powers of x. In favourable cases, this determines a set of coefficients and therefore an infinite series; the work of Chapter XXV then shows that the steps are justified for values of x lying within the radius of convergence of that series.

It may be helpful to remark (without proof) that if the given equation is written in the form

$$x^2\frac{d^2y}{dx^2}+xp(x)\frac{dy}{dx}+q(x)\,y=0,$$

and if the two functions $p(x)$, $q(x)$ can be expanded in series of ascending powers of x, then the solutions to be determined are convergent with radius of convergence equal to the lesser of the radii for $p(x)$ and $q(x)$. The reader will probably decide on brief reflection that it would be unlikely to be greater.

1. The method illustrated.

Consider the equation (of so-called 'hypergeometric' type)

$$(x-x^2)\,y''+(\tfrac{1}{2}-4x)\,y'-2y=0,$$

or

$$2(x-x^2)\,y''+(1-8x)\,y'-4y=0.$$

We seek solutions of the form

$$y=\sum_0^\infty a_\lambda x^{c+\lambda} \quad (a_0\neq 0;\text{ summation with respect to }\lambda),$$

for which

$$y'=\sum_0^\infty (c+\lambda)\,a_\lambda x^{c+\lambda-1},$$

$$y''=\sum_0^\infty (c+\lambda)(c+\lambda-1)\,a_\lambda x^{c+\lambda-2}.$$

Substitute in the given equation; then

$$\left\{2\sum_0^\infty (c+\lambda)(c+\lambda-1)\,a_\lambda x^{c+\lambda-1}-2\sum_0^\infty (c+\lambda)(c+\lambda-1)\,a_\lambda x^{c+\lambda}\right\}$$
$$+\left\{\sum_0^\infty (c+\lambda)\,a_\lambda x^{c+\lambda-1}-8\sum_0^\infty (c+\lambda)\,a_\lambda x^{c+\lambda}\right\}-4\sum_0^\infty a_\lambda x^{c+\lambda}\equiv 0,$$

or, grouping like summations,

$$\sum_0^\infty (c+\lambda)(2c+2\lambda-1)\,a_\lambda x^{c+\lambda-1}-2\sum_0^\infty (c+\lambda+2)(c+\lambda+1)\,a_\lambda x^{c+\lambda}\equiv 0.$$

The lowest power of x present, namely, x^{c-1}, occurs in the first summation only, when $\lambda=0$; for the vanishing of this term, we require

$$c(2c-1)\,a_0=0.$$

Hence, since $a_0 \neq 0$, c has one or other of the values

$$c=0 \quad \text{or} \quad c=\tfrac{1}{2}.$$

A general power, say x^{c+n}, occurs when $\lambda=n+1$ in the first summation and when $\lambda=n$ in the second; for the vanishing of this term, we require

$$(c+n+1)\,(2c+2n+1)\,a_{n+1}=2(c+n+2)\,(c+n+1)\,a_n.$$

[We must not yet cancel the factor $c+n+1$ from each side, lest it has significance.]

Taking $c=0, \frac{1}{2}$ in turn, we obtain the respective relations

$$\text{(i)} \quad (n+1)\,(2n+1)\,a_{n+1}=2(n+2)\,(n+1)\,a_n,$$

$$\text{(ii)} \quad (n+\tfrac{3}{2})\,(2n+2)\,a_{n+1}=2(n+\tfrac{5}{2})\,(n+\tfrac{3}{2})\,a_n.$$

Now $n \geqslant 0$ for all terms of the series, and so the factors $n+1$, $n+\frac{3}{2}$ are never zero. Hence the relations are

$$\text{(i)} \quad a_{n+1}=\frac{2(n+2)}{2n+1}\,a_n,$$

$$\text{(ii)} \quad a_{n+1}=\frac{2n+5}{2(n+1)}\,a_n.$$

We are therefore able to proceed to the evaluation of the two series corresponding to $c=0$, $c=\frac{1}{2}$ respectively.

(i) When $c=0$,

$$a_1=\frac{2.2}{1}\,a_0,$$

$$a_2=\frac{2.3}{3}\,a_1=\frac{2^2.2.3}{1.3}\,a_0,$$

$$a_3=\frac{2.4}{5}\,a_2=\frac{2^3.2.3.4}{1.3.5}\,a_0,$$

$$a_4=\frac{2.5}{7}\,a_3=\frac{2^4.2.3.4.5}{1.3.5.7}\,a_0,$$

and so on. Hence the series is

$$a_0\left\{1+\frac{2}{1}(2x)+\frac{2.3}{1.3}(2x)^2+\frac{2.3.4}{1.3.5}(2x)^3+\frac{2.3.4.5}{1.3.5.7}(2x)^4+\ldots\right\}.$$

(ii) When $c=\frac{1}{2}$, $a_1=\frac{5}{2}a_0$,

$$a_2=\frac{7}{2.2}a_1=\frac{5.7}{2^2.2}a_0,$$

$$a_3=\frac{9}{2.3}a_2=\frac{5.7.9}{2^3.2.3}a_0,$$

$$a_4=\frac{11}{2.4}a_3=\frac{5.7.9.11}{2^4.2.3.4}a_0,$$

and so on. Hence the series is

$$a_0x^{\frac{1}{2}}\left\{1+\frac{5}{1}\left(\frac{x}{2}\right)+\frac{5.7}{2!}\left(\frac{x}{2}\right)^2+\frac{5.7.9}{3!}\left(\frac{x}{2}\right)^3+\frac{5.7.9.11}{4!}\left(\frac{x}{2}\right)^4+\ldots\right\}.$$

If we denote the two series within brackets { } by the letters u, v respectively, then the general solution of the given equation, with two arbitrary constants, is

$$y=Au+Bx^{\frac{1}{2}}v.$$

The radius of convergence is easily obtained, since, for each series,

$$\lim_{n\to\infty}\left|\frac{a_n}{a_{n+1}}\right|=1,$$

so that

$$R=1.$$

Hence each series converges uniformly in any interval

$$|x|\leqslant 1-\delta \quad (\delta>0),$$

and the operations which we have performed are thus justified within that limitation.

NOTE. The equation, when expressed in the form

$$x^2y''+xp(x)\,y'+q(x)\,y=0,$$

is

$$x^2y''+x\left(\frac{1-8x}{2-2x}\right)y'+\left(\frac{-4x}{2-2x}\right)y=0,$$

and the (ordinary binomial) expansions for

$$\frac{1-8x}{2(1-x)}, \quad \frac{-4x}{2(1-x)}$$

both converge for $|x|<1$. Compare this result with the remark on p. 163.

The equation to determine c (p. 164) is called the INDICIAL EQUATION of the given equation, and its roots provide, in general, two starting-points for series to solve the given equation. There are, however, certain difficulties in particular problems, and we must now examine how they can arise.

2. Indicial equation with two equal roots. We illustrate the treatment of the equation when two roots of the indicial equation are equal by considering another equation of hypergeometric type,

$$(x-x^2)\,y''+(1-4x)\,y'-2y=0.$$

Write

$$y=\sum_0^\infty a_\lambda x^{c+\lambda} \quad (a_0 \neq 0),$$

so that

$$y'=\sum_0^\infty (c+\lambda)\,a_\lambda x^{c+\lambda-1},$$

$$y''=\sum_0^\infty (c+\lambda)\,(c+\lambda-1)\,a_\lambda x^{c+\lambda-2}.$$

Substitute in the given equation; then

$$\left\{\sum_0^\infty (c+\lambda)\,(c+\lambda-1)\,a_\lambda x^{c+\lambda-1}-\sum_0^\infty (c+\lambda)\,(c+\lambda-1)\,a_\lambda x^{c+\lambda}\right\}$$
$$+\left\{\sum_0^\infty (c+\lambda)a_\lambda x^{c+\lambda-1}-4\sum_0^\infty (c+\lambda)\,a_\lambda x^{c+\lambda}\right\}-2\sum_0^\infty a_\lambda x^{c+\lambda}\equiv 0,$$

or, grouping like summations,

$$\sum_0^\infty (c+\lambda)^2\,a_\lambda x^{c+\lambda-1}-\sum_0^\infty (c+\lambda+1)\,(c+\lambda+2)\,a_\lambda x^{c+\lambda}\equiv 0.$$

The lowest power of x present, namely, x^{c-1}, occurs in the first summation only, when $\lambda=0$; thus

$$c^2 a_0=0 \quad (a_0 \neq 0).$$

The two roots of this equation for c are equal, namely,

$$c=0,0.$$

A general power, say x^{c+n} (for $n \geqslant 0$), occurs when $\lambda = n+1$ in the first summation and when $\lambda = n$ in the second; thus

$$(c+n+1)^2 a_{n+1} = (c+n+1)(c+n+2) a_n.$$

Since we have proved that $c=0$, the factor $c+n+1$ cannot be zero for $n \geqslant 0$, and may therefore be cancelled. Hence

$$a_{n+1} = \frac{c+n+2}{c+n+1} a_n,$$

or, with $c=0$,
$$a_{n+1} = \frac{n+2}{n+1} a_n.$$

Thus
$$a_1 = \tfrac{2}{1} a_0,$$
$$a_2 = \tfrac{3}{2} a_1 = 3a_0,$$
$$a_3 = \tfrac{4}{3} a_2 = 4a_0,$$
$$a_4 = \tfrac{5}{4} a_3 = 5a_0,$$

and so on. Hence the series is

$$a_0\{1 + 2x + 3x^2 + 4x^3 + \dots\},$$

valid for $|x| < 1$.

[In this particular case, we recognize the solution in finite terms as $a_0(1-x)^{-2}$.]

The new feature of this discussion, as compared with that in § 1, is that, because of the *repeated* root $c=0$, we have (so far) been able to obtain only ONE solution of the given differential equation. We must therefore undertake further investigations for a second solution.

We begin by forming the solution in series as before, *but ignoring the fact that* $c=0$. Then

$$a_1 = \frac{c+2}{c+1} a_0,$$

$$a_2 = \frac{c+3}{c+2} a_1,$$

$$a_3 = \frac{c+4}{c+3} a_2,$$

and so on. These coefficients satisfy the relation

$$(c+n+1)^2 a_{n+1} = (c+n+1)(c+n+2) a_n$$

for all values of n such that $n \geqslant 0$, and so it follows that, when this series is substituted in the given equation, all terms vanish except that in x^{c-1}, which gives (as before) c^2a_0. Hence, if we write

$$u \equiv a_0\left\{x^c + \frac{c+2}{c+1}x^{c+1} + \frac{c+3}{c+1}x^{c+2} + \frac{c+4}{c+1}x^{c+3} + \dots\right\},$$

then, identically,

$$(x-x^2)\,u'' + (1-4x)\,u' - 2u \equiv a_0 c^2 x^{c-1}.$$

This relation holds, subject to convergence, for any value of c.

Now repeated roots are associated, almost by tradition, with the vanishing of a differential coefficient, and we observe here that the differential coefficient, with respect to c, of the right-hand side vanishes when $c = 0$. Hence the differential coefficient of the left-hand side also vanishes, so that

$$\frac{\partial}{\partial c}\{(x-x^2)\,u'' + (1-4x)\,u' - 2u\}_{c=0} = 0,$$

where c is equated to zero *after* the differentiation.

Finally, we establish two 'commutability' relations

$$\frac{\partial}{\partial c}\left(\frac{du}{dx}\right) = \frac{d}{dx}\left(\frac{\partial u}{\partial c}\right), \quad \frac{\partial}{\partial c}\left(\frac{d^2u}{dx^2}\right) = \frac{d^2}{dx^2}\left(\frac{\partial u}{\partial c}\right).$$

To prove them, consider each term $a_n x^{n+c}$ separately. We have, by logarithmic differentiation, the formula

$$\frac{\partial}{\partial c}(a_n x^{n+c}) = x^{n+c}\frac{\partial a_n}{\partial c} + a_n x^{n+c}\log x,$$

so that

$$\frac{d}{dx}\left\{\frac{\partial}{\partial c}(a_n x^{n+c})\right\} = (n+c)\,x^{n+c-1}\frac{\partial a_n}{\partial c} + a_n x^{n+c-1} + (n+c)\,a_n x^{n+c-1}\log x;$$

also $\quad \dfrac{d}{dx}(a_n x^{n+c}) = (n+c)\,a_n x^{n+c-1},$

so that

$$\frac{\partial}{\partial c}\left\{\frac{d}{dx}(a_n x^{n+c})\right\} = (n+c)\,x^{n+c-1}\frac{\partial a_n}{\partial c} + a_n x^{n+c-1} + (n+c)\,a_n x^{n+c-1}\log x.$$

Hence the first relation holds for each individual term, and so for the whole series, within its radius of convergence. But see the Note at the end of this section.

Moreover

$$\frac{\partial}{\partial c}\left(\frac{d^2u}{dx^2}\right)=\frac{\partial}{\partial c}\left(\frac{du'}{dx}\right)$$

$$=\frac{d}{dx}\left(\frac{\partial u'}{\partial c}\right)$$

$$=\frac{d}{dx}\left\{\frac{d}{dx}\left(\frac{\partial u}{\partial c}\right)\right\}$$

$$=\frac{d^2}{dx^2}\left(\frac{\partial u}{\partial c}\right).$$

The formulae of commutability are therefore established.

Returning to the relation

$$\frac{\partial}{\partial c}\{(x-x^2)\,u''+(1-4x)\,u'-2u\}_{c=0}=0,$$

we are now able to express it in the form

$$(x-x^2)\frac{d^2}{dx^2}\left(\frac{\partial u}{\partial c}\right)_0+(1-4x)\frac{d}{dx}\left(\frac{\partial u}{\partial c}\right)_0-2\left(\frac{\partial u}{\partial c}\right)_0=0,$$

so that *there exists a second solution of the given equation, in the form*

$$\left(\frac{\partial u}{\partial c}\right)_0.$$

But a typical term of the series for u is

$$\frac{c+n+1}{c+1}x^{c+n},$$

and $$\frac{\partial}{\partial c}\left(\frac{c+n+1}{c+1}x^{c+n}\right)=\frac{c+n+1}{c+1}x^{c+n}\log x-\frac{n}{(c+1)^2}x^{c+n}.$$

Hence $$\frac{\partial}{\partial c}\left(\frac{c+n+1}{c+1}x^{c+n}\right)_0=(n+1)\,x^n\log x-nx^n.$$

The corresponding solution is therefore

$$(1+2x+3x^2+4x^3+\dots)\log x-(x+2x^2+3x^3+\dots).$$

The complete solution of the given equation is thus

$$y=A(1+2x+3x^2+\dots)$$
$$+B\{(1+2x+3x^2+\dots)\log x-(x+2x^2+3x^3+\dots)\},$$

where A, B are arbitrary constants.

Each of the infinite series is convergent for $|x| < 1$, though, of course, $\log x$ is meaningless when $x = 0$.

NOTE. We have glossed over the difficulty that the statement

$$\text{`}\frac{\partial}{\partial c}\sum^{\infty} a_n(c)\,x^{n+c} = \sum^{\infty}\frac{\partial}{\partial c}\{a_n(c)\,x^{n+c}\}\text{'}$$

requires proof. This is hard, and cannot be attempted here. On the other hand, we may use the argument that the above analysis indicates what a solution of the equation might be; and, *for the particular numbers*, substitution of the proposed solution into the given equation verifies that the supposition is correct.

3. Indicial equation with roots differing by an integer (leading to 'infinite coefficients'). We turn to another equation of hypergeometric type,

$$(x - x^2)\,y'' - 4xy' - 2y = 0.$$

Writing

$$y = \sum_0^{\infty} a_\lambda x^{c+\lambda}$$

and proceeding exactly as in the two preceding sections, we reach the equations

$$c(c-1)\,a_0 = 0 \quad (a_0 \neq 0),$$

$$(c+n+1)\,(c+n)\,a_{n+1} = (c+n+1)\,(c+n+2)\,a_n.$$

Thus

$$c = 0 \quad \text{or} \quad c = 1;$$

also, since

$$c+n+1 \neq 0 \quad \text{for} \quad n \geqslant 0,$$

$$(c+n)\,a_{n+1} = (c+n+2)\,a_n.$$

The new feature of this example is that the two roots $c = 0$, $c = 1$ differ by an integer.

The successive coefficients are calculated according to the sequence

$$a_1 = \frac{c+2}{c}\,a_0,$$

$$a_2 = \frac{c+3}{c+1}\,a_1 = \frac{(c+2)\,(c+3)}{c(c+1)}\,a_0,$$

$$a_3 = \frac{c+4}{c+2}\,a_2 = \frac{(c+3)\,(c+4)}{c(c+1)}\,a_0,$$

$$a_4 = \frac{c+5}{c+3}\,a_3 = \frac{(c+4)\,(c+5)}{c(c+1)}\,a_0,$$

and so on.

The solution $c=1$ is straightforward, and gives the series

$$a_0\left\{x+3x^2+\frac{3.4}{2}x^3+\frac{4.5}{2}x^4+\dots\right\},$$

which, in fact, we recognize as $a_0x(1-x)^{-3}$.

The solution $c=0$ seems at first sight to be useless, since the equation $ca_1=(c+2)a_0$ cannot be satisfied for any value of a_1 when $a_0 \neq 0$. We find, however, that a solution may be obtained in the following way:

Beginning with the general form

$$a_0\left\{x^c+\frac{c+2}{c}x^{c+1}+\frac{(c+2)(c+3)}{c(c+1)}x^{c+2}+\frac{(c+3)(c+4)}{c(c+1)}x^{c+3}+\dots\right\},$$

we can get rid of the factor c from the denominators by writing a_0 in the form kc (with $k \neq 0$). This device introduces a value of a_0 which vanishes with c, but we shall see that this disadvantage does not prove fatal.

Putting $$a_0 \equiv kc \quad (k \neq 0),$$
we write

$$u \equiv k\left\{cx^c+(c+2)x^{c+1}+\frac{(c+2)(c+3)}{c+1}x^{c+2}+\frac{(c+3)(c+4)}{c+1}x^{c+3}+\dots\right\}.$$

Substituting this expression in the given equation, we find the relation

$$(x-x^2)u''-4xu'-2u \equiv kc^2(c-1)x^{c-1}.$$

Since

$$\frac{\partial}{\partial c}\{kc^2(c-1)x^{c-1}\}_{c=0}=0,$$

it follows, exactly as in the preceding section, that

$$(x-x^2)\frac{d^2}{dx^2}\left(\frac{\partial u}{\partial c}\right)_0-4x\frac{d}{dx}\left(\frac{\partial u}{\partial c}\right)_0-2\left(\frac{\partial u}{\partial c}\right)_0=0,$$

so that *a second solution of the given equation is*

$$\left(\frac{\partial u}{\partial c}\right)_{c=0}.$$

In order to evaluate this expression, note that

$$\frac{\partial}{\partial c}(x^{c+n})=x^{c+n}\log x,$$

so that

$$\frac{\partial}{\partial c}(x^{c+n})_0=x^n\log x.$$

Also, since, for small values of c,

$$\frac{(c+n)(c+n+1)}{c+1} = (c+n)(c+n+1)(1-c+c^2-\dots)$$
$$= n(n+1) - (n^2-n-1)c + \dots,$$
$$\frac{\partial}{\partial c}\left\{\frac{(c+n)(c+n+1)}{c+1}\right\}_0 = -(n^2-n-1).$$

We therefore have the solution

$$u = k\{(2x + 2.3x^2 + 3.4x^3 + \dots)\log x$$
$$+ (1 + x - x^2 - 5x^3 - 11x^4 - \dots)\}.$$

The general solution of the given equation is thus

$$y = A[1.2x + 2.3x^2 + 3.4x^3 + \dots + n(n+1)x^n + \dots]$$
$$+ B[\{1.2x + 2.3x^2 + \dots + n(n+1)x^n + \dots\}\log x$$
$$+ \{1 + x - x^2 - 5x^3 - \dots - (n^2-n-1)x^n - \dots\}].$$

There is one final point to be cleared. Not only does $(\partial u/\partial c)_0$ satisfy the given equation, but so also, and more obviously, does u_0 itself. But this merely gives

$$k\{2x + 2.3x^2 + 3.4x^3 + \dots\},$$

and so provides nothing new.

4. Indicial equation with roots differing by an integer (leading to indeterminate coefficients). Still keeping to the hypergeometric type, consider the equation

$$(x - x^2)y'' - (1+2x)y' + 2y = 0.$$

Writing
$$y = \sum_0^\infty a_\lambda x^{c+\lambda}$$

and proceeding as before, we obtain the equations

$$c(c-2)a_0 = 0 \quad (a_0 \neq 0),$$
$$(c+n+1)(c+n-1)a_{n+1} = (c+n+2)(c+n-1)a_n.$$

Thus
$$c = 0 \quad \text{or} \quad c = 2.$$

The new feature in this example is the possibility that the factor $c+n-1$ can vanish on *each* side when $c=0$ (for the value $n=1$).

The solution for $c=2$ is straightforward, since then

$$a_{n+1}=\frac{n+4}{n+3}a_n,$$

so that

$$a_1=\tfrac{4}{3}a_0,$$

$$a_2=\tfrac{5}{4}a_1=\tfrac{5}{3}a_0,$$

$$a_3=\tfrac{6}{5}a_2=\tfrac{6}{3}a_0,$$

and so on. The series is

$$a_0\{x^2+\tfrac{4}{3}x^3+\tfrac{5}{3}x^4+\ldots\},$$

which, in fact, we recognize as

$$\tfrac{1}{3}a_0\{(1-x)^{-2}-(1+2x)\}.$$

When $c=0$, the relation connecting successive terms is

$$(n+1)(n-1)a_{n+1}=(n+2)(n-1)a_n.$$

For $n=0$, we have
$$-a_1=-2a_0,$$
or
$$a_1=2a_0.$$

For $n=1$ we have no information, since each side of the relation vanishes identically, being satisfied whatever a_2 may be.

For $n\geqslant 2$, we can cancel the factor $n-1$, thereby obtaining the relation

$$(n+1)a_{n+1}=(n+2)a_n \quad (n\geqslant 2).$$

The coefficient a_2 may be given an *arbitrary* value, and then we have

$$a_3=\tfrac{4}{3}a_2,$$

$$a_4=\tfrac{5}{4}a_3,$$

and so on.

We therefore obtain for $c=0$ the series

$$a_0(1+2x)+a_2(x^2+\tfrac{4}{3}x^3+\tfrac{5}{3}x^4+\ldots),$$

and we observe that the coefficient of a_2 is just the series already obtained for $c=2$.

Hence we have in all the general solution

$$y=A(1+2x)+B(3x^2+4x^3+5x^4+\ldots).$$

See also another example of indeterminate coefficients in connection with the solution of *Legendre's equation* on p. 174.

5. The hypergeometric series. We shall not be dealing directly with the important series known as the HYPERGEOMETRIC SERIES, but, as we have used a number of particular examples in illustration, we give the general definition of the series:

$$1+\frac{\alpha\beta}{1.\gamma}x+\frac{\alpha(\alpha+1)\beta(\beta+1)}{1.2.\gamma(\gamma+1)}x^2$$
$$+\frac{\alpha(\alpha+1)(\alpha+2)\beta(\beta+1)(\beta+2)}{1.2.3.\gamma(\gamma+1)(\gamma+2)}x^3+\ldots,$$

convergent if $|x|<1$. It is a solution of the HYPERGEOMETRIC EQUATION

$$x(1-x)y''+\{\gamma-(\alpha+\beta+1)x\}y'-\alpha\beta y=0.$$

The series is often denoted by the notation

$$F(\alpha,\beta,\gamma,x).$$

It is easy to verify that the indicial equation for solutions in the form

$$y=\sum_0^\infty a_\lambda x^{c+\lambda}$$

has roots

$$c=0, \quad c=1-\gamma,$$

so that, if γ is not an integer (or zero), a second series satisfying the hypergeometric equation is

$$x^{1-\gamma}F(\alpha-\gamma+1,\beta-\gamma+1,2-\gamma,x).$$

6. Legendre's equation. Another important equation soluble in series is LEGENDRE'S EQUATION

$$(1-x^2)y''-2xy'+n(n+1)y=0,$$

about which we shall have more to say later. In the meantime we try, as usual, the series

$$y=\sum_0^\infty a_\lambda x^{c+\lambda},$$

so that

$$y'=\sum_0^\infty (c+\lambda)a_\lambda x^{c+\lambda-1},$$

$$y''=\sum_0^\infty (c+\lambda)(c+\lambda-1)a_\lambda x^{c+\lambda-2}.$$

Substitute in the given equation; then

$$\sum_{0}^{\infty} (c+\lambda)(c+\lambda-1)\, a_\lambda x^{c+\lambda-2}$$
$$-\sum_{0}^{\infty} \{(c+\lambda)(c+\lambda-1)+2(c+\lambda)-n(n+1)\}\, a_\lambda x^{c+\lambda}=0.$$

The lowest power of x present, namely, x^{c-2}, occurs in the first summation only, when $\lambda=0$; thus

$$c(c-1)\, a_0=0.$$

The next power, that of x^{c-1}, again occurs in the first summation only, when $\lambda=1$; thus

$$(c+1)\, ca_1=0.$$

A general power, say x^{c+r} (for $r \geqslant 0$), occurs when $\lambda=r+2$ in the first summation and when $\lambda=r$ in the second; thus

$$(c+r+2)(c+r+1)\, a_{r+2}=(c+r+n+1)(c+r-n)\, a_r.$$

From the indicial equation we obtain the two possibilities

$$c=0 \quad \text{or} \quad c=1,$$

which we consider in turn.

(i) *The solution* $c=0$. When $c=0$ the equation $(c+1)\, ca_1=0$ is satisfied automatically, and so the coefficient a_1 is indeterminate. For the succeeding coefficients the relation is

$$(r+2)(r+1)\, a_{r+2}=(r+n+1)(r-n)\, a_r \quad (r \geqslant 0),$$

or

$$a_{r+2}=\frac{(r+n+1)(r-n)}{(r+2)(r+1)}\, a_r.$$

The 'even' coefficients may now be calculated in succession:

$$a_2=\frac{(n+1)(-n)}{2!}\, a_0,$$

$$a_4=\frac{(n+3)(2-n)}{4.3}\, a_2=\frac{(n+3)(n+1)(2-n)(-n)}{4!}\, a_0,$$

$$a_6=\frac{(n+5)(4-n)}{6.5}\, a_4=\frac{(n+5)(n+3)(n+1)(4-n)(2-n)(-n)}{6!}\, a_0,$$

and so on.

Similarly for the 'odd' coefficients,

$$a_3 = \frac{(n+2)(1-n)}{3!} a_1,$$

$$a_5 = \frac{(n+4)(3-n)}{5.4} a_3 = \frac{(n+4)(n+2)(3-n)(1-n)}{5!} a_1,$$

$$a_7 = \frac{(n+6)(5-n)}{7.6} a_5 = \frac{(n+6)(n+4)(n+2)(5-n)(3-n)(1-n)}{7!} a_1,$$

and so on.

The general solution of Legendre's equation is therefore obtained in the form

$$y = a_0\left\{1 + \frac{(n+1)(-n)}{2!} x^2 + \frac{(n+3)(n+1)(2-n)(-n)}{4!} x^4 + \ldots\right\}$$

$$+ a_1\left\{x + \frac{(n+2)(1-n)}{3!} x^3 + \frac{(n+4)(n+2)(3-n)(1-n)}{5!} x^5 + \ldots\right\}.$$

It is easy to verify that each of the two series in this expression converges for $|x| < 1$.

(ii) *The solution* $c = 1$. This condition gives us nothing essentially new. The relation $(c+1)ca_1 = 0$ gives $a_1 = 0$, so that, by the recurrence relation,

$$a_1 = a_3 = a_5 = \ldots = a_{2r+1} = \ldots = 0.$$

Further $\qquad (r+3)(r+2)a_{r+2} = (r+n+2)(r+1-n)a_r,$

so that

$$a_2 = \frac{(n+2)(1-n)}{3!} a_0,$$

$$a_4 = \frac{(n+4)(3-n)}{5.4} a_2 = \frac{(n+4)(n+2)(3-n)(1-n)}{5!} a_0,$$

and so on. Thus the series is

$$a_0\left\{x + \frac{(n+2)(1-n)}{3!} x^3 + \frac{(n+4)(n+2)(3-n)(1-n)}{5!} x^5 + \ldots\right\},$$

a series identical with the coefficient of a_1 in the general solution obtained previously.

COROLLARY. When n is an even integer, there is a solution (the coefficient of a_0 in the general solution) which is a polynomial of order n whose terms are even powers of x; when n is an odd integer, there is a solution (the coefficient of a_1 in the general solution) which is a polynomial of order n whose terms are odd powers of x. These sets of solutions will be unified and assume importance later.

7. Bessel's equation. The equation

$$x^2y'' + xy' + (x^2 - n^2)\,y = 0,$$

known as BESSEL'S EQUATION, will also arise in the subsequent work. To solve it, we begin with the series

$$y = \sum_0^\infty a_\lambda x^{c+\lambda},$$

from which we obtain, by substitution in the given equation, the relation

$$\sum_0^\infty \{(c+\lambda)(c+\lambda-1) + (c+\lambda) - n^2\}\, a_\lambda x^{c+\lambda} + \sum_0^\infty a_\lambda x^{c+\lambda+2} = 0,$$

or

$$\sum_0^\infty (c+\lambda+n)(c+\lambda-n)\, a_\lambda x^{c+\lambda} + \sum_0^\infty a_\lambda x^{c+\lambda+2} = 0.$$

The lowest power of x present, namely, x^c, occurs in the first summation, when $\lambda = 0$; thus

$$(c+n)(c-n)\,a_0 = 0.$$

The next power, that of x^{c+1}, again occurs in the first summation only, when $\lambda = 1$; thus

$$(c+1+n)(c+1-n)\,a_1 = 0.$$

A general power, say x^{c+r} (for $r \geqslant 2$), occurs when $\lambda = r$ in the first summation and when $\lambda = r-2$ in the second; thus

$$(c+r+n)(c+r-n)\,a_r + a_{r-2} = 0,$$

or, replacing r by $r+2$,

$$(c+r+2+n)(c+r+2-n)\,a_{r+2} = -a_r \quad (r \geqslant 0).$$

From the indicial equation we obtain the two possibilities

$$c = n \quad \text{or} \quad c = -n,$$

though these give the samo value when $n = 0$.

The coefficient of a_1 in the relation found above is

$$(c+1+n)(c+1-n).$$

For $c=n$, that is $\qquad 2n+1,$

and for $c=-n$, it is $\qquad 1-2n.$

Hence a_1 is indeterminate (since the product of factors vanishes) for $n=\pm\frac{1}{2}$, but is otherwise zero.

Suppose that $n\neq\pm\frac{1}{2}$, so that $a_1=0$. Then, by the recurrence relation, we have the relations

$$a_1=a_3=a_5=\ldots=0,$$

and so the series contains only the 'even' terms. Writing $r\equiv 2s$, we obtain the relation between successive coefficients in the form

$$a_{2s+2}=\frac{-a_{2s}}{(c+2s+2+n)(c+2s+2-n)} \quad (s\geqslant 0),$$

where $c=n$ or $c=-n$.

Consider the solution for $c=n$. The relation is

$$a_{2s+2}=\frac{-a_{2s}}{4(n+1+s)(s+1)}.$$

If we exclude the case when n is a negative integer, we obtain the coefficients in the form

$$a_2=\frac{-a_0}{4(n+1).1},$$

$$a_4=\frac{-a_2}{4(n+2).2}=\frac{a_0}{4^2(n+1)(n+2).2!},$$

$$a_6=\frac{-a_4}{4(n+3).3}=\frac{-a_0}{4^3(n+1)(n+2)(n+3).3!},$$

and so on. Hence we obtain the series

$$x^n\left\{1-\frac{1}{(n+1)}\left(\frac{x}{2}\right)^2+\frac{1}{(n+1)(n+2).2!}\left(\frac{x}{2}\right)^4 -\frac{1}{(n+1)(n+2)(n+3).3!}\left(\frac{x}{2}\right)^6+\ldots\right\}.$$

Consider next the solution for $c=-n$. The argument is exactly the same as that just given for $c=n$, save that now n must not be a positive integer, and we obtain the series

$$x^{-n}\left\{1-\frac{1}{(-n+1)}\left(\frac{x}{2}\right)^2+\frac{1}{(-n+1)(-n+2).2!}\left(\frac{x}{2}\right)^4 - \frac{1}{(-n+1)(-n+2)(-n+3).3!}\left(\frac{x}{2}\right)^6+\ldots\right\}.$$

Hence, when n is not zero, $\pm\frac{1}{2}$, or an integer, a general solution of Bessel's equation is obtained in the form

$$y=Ax^n\left\{1-\frac{1}{(n+1)}\left(\frac{x}{2}\right)^2+\frac{1}{(n+1)(n+2).2!}\left(\frac{x}{2}\right)^4-\ldots\right\} + Bx^{-n}\left\{1-\frac{1}{(-n+1)}\left(\frac{x}{2}\right)^2+\frac{1}{(-n+1)(-n+2).2!}\left(\frac{x}{2}\right)^4-\ldots\right\}.$$

The two series converge for all values of x.

When n is a negative integer, A must be taken as zero; when n is a positive integer, B must be taken as zero. When n is zero, the two series are identical. In each of these cases, a further solution is required and may be obtained if required by the methods indicated for the equations of hypergeometric type earlier in this chapter. We do not make any detailed study here.

The anomaly which we mentioned for $n=\pm\frac{1}{2}$ is apparent rather than real, for then the series arising from a_0 and the series arising from a_1 turn out to be just the two series which we have already obtained.

EXAMPLES I

Find the general solution, in series of ascending powers of x, of the equations:

1. $x^2y''+xy'+(x^2-\frac{9}{4})y=0.$
2. $xy''+y'+xy=0.$
3. $x^2y''+xy'+(x^2-1)y=0.$
4. $x^2y''+xy'+(x^2-\frac{1}{4})y=0.$

8. Solution in descending powers of x; Legendre's equation. Sometimes a solution is required which expresses y in a series of *descending* powers of x. We take as an example Legendre's equation, for which this approach turns out to be particularly convenient.

The equation is

$$(1-x^2)\,y''-2xy'+n(n+1)\,y=0,$$

and we assume the existence of a solution in the form

$$y=\sum_0^\infty b_\lambda\,x^{c-\lambda},$$

so that

$$y'=\sum_0^\infty (c-\lambda)\,b_\lambda x^{c-\lambda-1},$$

$$y''=\sum_0^\infty (c-\lambda)\,(c-\lambda-1)\,b_\lambda x^{c-\lambda-2}.$$

Substitute in the given equation; then

$$\left\{\sum_0^\infty (c-\lambda)\,(c-\lambda-1)\,b_\lambda x^{c-\lambda-2}-\sum_0^\infty (c-\lambda)\,(c-\lambda-1)\,b_\lambda x^{c-\lambda}\right\}$$
$$-2\sum_0^\infty (c-\lambda)\,b_\lambda x^{c-\lambda}+n(n+1)\sum_0^\infty b_\lambda x^{c-\lambda}=0,$$

or

$$\sum_0^\infty (c-\lambda)\,(c-\lambda-1)\,b_\lambda x^{c-\lambda-2}$$
$$-\sum_0^\infty \{(c-\lambda)\,(c-\lambda+1)-n(n+1)\}\,b_\lambda x^{c-\lambda}=0.$$

The highest power of x present, namely, x^c, occurs in the second summation only, when $\lambda=0$; thus

$$\{c(c+1)-n(n+1)\}\,b_0=0,$$

or

$$(c-n)\,(c+n+1)=0.$$

The next power, that of x^{c-1}, again occurs in the second summation only, when $\lambda=1$; thus

$$(c+n)\,(c-n-1)\,b_1=0.$$

A general power, say x^{c-r} (for $r\geqslant 2$), occurs when $\lambda=r-2$ in the first summation and when $\lambda=r$ in the second; thus

$$(c-r+2)\,(c-r+1)\,b_{r-2}=\{(c-r)\,(c-r+1)-n(n+1)\}\,b_r$$
$$=(c-r-n)\,(c-r+n+1)\,b_r.$$

The indicial equation gives the two possibilities

$$c=n,\quad c=-(n+1),$$

which we must consider in turn.

(i) *The solution* $c=n$. When $c=n$, the second equation just obtained is

$$-2nb_1=0,$$

and so (assuming $n\neq 0$) $\quad b_1=0.$

The general equation is

$$(n-r+2)(n-r+1)b_{r-2}=-r(2n-r+1)b_r,$$

or

$$b_r=-\frac{(n-r+2)(n-r+1)}{r(2n-r+1)}b_{r-2}.$$

Since $b_1=0$, it follows that

$$b_3=b_5=b_7=\ldots=0.$$

Also,

$$b_2=-\frac{n(n-1)}{2(2n-1)}b_0,$$

$$b_4=-\frac{(n-2)(n-3)}{4(2n-3)}b_2=\frac{n(n-1)(n-2)(n-3)}{2.4.(2n-1)(2n-3)}b_0,$$

and so on. Hence the series is

$$b_0x^n\left\{1+\frac{n(n-1)}{(2n-1)}\left(\frac{-1}{2x^2}\right)+\frac{n(n-1)(n-2)(n-3)}{2!(2n-1)(2n-3)}\left(\frac{-1}{2x^2}\right)^2+\ldots\right\}.$$

The case '$2n$=odd integer' requires further consideration, which we do not give here. On the other hand, we shall have occasion later to refer to the case

$$n=\text{positive integer},$$

for which the series terminates, giving a solution which is a polynomial in x of degree n. This is, in fact, the same solution that we obtained earlier (p. 176) in a form that appeared different according as n was odd or even.

(ii) *The solution* $c=-(n+1)$. When $c=-(n+1)$, the equation involving b_1 is

$$2(n+1)b_1=0,$$

and so (assuming $n\neq -1$) $\quad b_1=0.$

The general equation is

$$(-n-r+1)(-n-r)b_{r-2}=(-2n-r-1)(-r)b_r,$$

or

$$b_r=\frac{(n+r)(n+r-1)}{r(2n+r+1)}b_{r-2}.$$

Since $b_1 = 0$, it follows that

$$b_3 = b_5 = b_7 = \ldots = 0.$$

Also, $$b_2 = \frac{(n+2)(n+1)}{2(2n+3)} b_0,$$

$$b_4 = \frac{(n+4)(n+3)}{4(2n+5)} b_2 = \frac{(n+4)(n+3)(n+2)(n+1)}{2.4(2n+5)(2n+3)} b_0,$$

and so on. Hence the series is

$$\frac{b_0}{x^{n+1}}\left\{1 + \frac{(n+1)(n+2)}{2n+3}\left(\frac{1}{2x^2}\right) + \frac{(n+1)(n+2)(n+3)(n+4)}{2!(2n+3)(2n+5)}\left(\frac{1}{2x^2}\right)^2 + \ldots\right\}.$$

The general solution of Legendre's equation is obtained by adding constant multiples of these two series, with possible examination of special cases.

For future reference, we repeat the result that, *if n is a positive integer, there is a solution which is a polynomial of order n in x, namely,*

$$y = A\left\{x^n + \frac{n(n-1)}{(2n-1)}\left(\frac{-1}{2}\right)x^{n-2} + \frac{n(n-1)(n-2)(n-3)}{2!(2n-1)(2n-3)}\left(\frac{-1}{2}\right)^2 x^{n-4} + \ldots\right\}$$

$$= \frac{(n!)^2}{(2n)!} A \Sigma \frac{(-1)^r (2n-2r)!}{r!(n-r)!(n-2r)!} x^{n-2r}.$$

REVISION EXAMPLES XXII

1. Prove that the equation

$$3x^2y'' + (3x^2 - x)y' + (1 - 9x)y = 0$$

has a solution of the form

$$y = Axy_1 + Bx^{\frac{1}{3}}y_2,$$

where y_1 is a quadratic function of x, and y_2 is a power series

$$y_2 \equiv 1 + 8x + 5x^2 + \ldots.$$

Find the general term in y_2.

2. Obtain a solution of the equation

$$x^2y'' + xy' + (x^2 - n^2)\,y = 0$$

in the form of a power series y_1, where n is a positive integer. Show, by putting $y = y_1 z$, that the general solution is

$$y = y_1\left\{A + B\int \frac{dx}{y_1^2 x}\right\}.$$

3. Find the general solution in series of the equation

$$2x(1-x^2)\frac{d^2y}{dx^2} + \frac{dy}{dx} + 12xy = 0.$$

Determine the range of values of x for which the solution is valid.

4. Show that the equation

$$2xy'' + (1-2x)\,y' - y = 0$$

has a solution of the form

$$y = Af(x) + Bx^{\frac{1}{2}}g(x),$$

where A, B are arbitrary constants and $f(x)$, $g(x)$ are series each of the form (to be found)

$$1 + a_1x + a_2x^2 + \ldots.$$

By putting $y = ve^x$ in the given equation, show that

$$\frac{d}{dx}\{e^{-x}x^{\frac{1}{2}}g(x)\} = \tfrac{1}{2}e^{-x}x^{-\frac{1}{2}}.$$

5. Obtain the solution

$$A\sum_{n=0}^{N} a_nx^n + Bx^{\frac{1}{2}}\sum_{n=0}^{\infty} b_nx^n$$

of the differential equation

$$2x(1-x)\,y'' + (1-x)\,y' + y = 0.$$

Show that the radius of convergence of Σb_nx^n is 1.

6. By solving in series, or otherwise, show that every solution of the differential equation

$$x^2(x+3)\,y''' - 3x(x+2)\,y'' + 6(x+1)\,y' - 6y = 0$$

is a polynomial in x.

7. Find the complete solution in series of the equation

$$x(1+2x^2)y''+2y'-12xy=0,$$

and give the range of values of x for which it is valid.

8. Find a solution as a power series of the equation

$$x(x-1)y''+3xy'+y=0,$$

and state where the series converges.

Identify the rational function of x represented by the series, and derive a second independent solution of the differential equation.

9. Integrate the equation

$$xy''+ky'-y=0$$

by the method of solution in series (i) when the constant k is not an integer, (ii) when $k=1$.

Express the general solution in finite form when $k=\frac{1}{2}$.

10. Obtain in the form of an infinite series the general solution of the differential equation

$$xy''+(k-x)y'-y=0.$$

Show that, if k is an integer greater than 1, the equation is satisfied by a polynomial of degree $k-1$ in x^{-1}.

11. By putting $y=x^c(a_0+a_1x+a_2x^2+\ldots)$, solve the equation

$$x^2(x^2-1)y''+(5x^3-4x)y'+(3x^2-2)y=0.$$

Show that one solution takes a very simple finite form.

12. Show that the equation

$$xy''-(x+4)y'+2y=0$$

has solutions (to be found) of which one is a quadratic expression and another a power series convergent for all values of x.

13. Find two distinct solutions in ascending powers of x of the equation

$$xy''-2y'+9x^5y=0.$$

14. Obtain the solution in series of the differential equation

$$(x^2+1)y''-3xy'+\lambda y=0.$$

Find all the values of λ for which the equation has a polynomial solution, and find all such solutions.

15. Find the general solution of the differential equation

$$xy'' + (x+1)\,y' + 3y = 0$$

in a form involving power series in x.

Find also the particular solution of the form

$$y = 1 + \sum_{n=1}^{\infty} a_n x^n,$$

and express this in finite terms.

16. Obtain the complete solution of the differential equation

$$x(1-x)\,y'' + (1-3x)\,y' - y = 0$$

in the form of a series convergent for small values of x.

17. By means of infinite series, solve the differential equation

$$(x^2-2x)\,y'' + 6y' - 6y = 0.$$

18. Prove that the differential equation

$$(1-x^2)\,y'' - 2xy' + 72y = 0$$

has a solution which is a polynomial in x, and find a second solution in series of powers of x.

19. Show that the differential equation

$$x^2y'' + (2x+x^2)\,y' + \{x - k(k+1)\}\,y = 0,$$

where k is constant and $2k$ is not an integer, has the two formal solutions

$$\sum_0^{\infty} (-1)^n \frac{(k+1)(k+2)\dots(k+n)}{(2k+2)(2k+3)\dots(2k+n+1)} \frac{x^{n+k}}{n!},$$

$$\sum_0^{\infty} (-1)^n \frac{k(k-1)\dots(k-n+1)}{2k(2k-1)\dots(2k-n+1)} \frac{x^{n-k-1}}{n!}.$$

20. Obtain two independent solutions in series of the equation

$$x^2(1-x)\,y'' - x(1+3x)\,y' + (1-x)\,y = 0.$$

Find the radii of convergence of the series.

CHAPTER XXVIII

FOURIER SERIES

1. Introductory example; vibrating string. Suppose that an elastic string, stretched between two fixed points O, A at a distance l apart, makes small vibrations in a plane containing those points in such a way that the elements of the string move at right angles to its length. It is known that the displacement y at time t of the point distant x from O satisfies the partial differential equation

$$\frac{\partial^2 y}{\partial t^2} = a^2 \frac{\partial^2 y}{\partial x^2},$$

where a is a constant depending on the nature of the string.

Solutions of this equation may be obtained in many ways. Our immediate problem is to investigate those solutions which assume the form

$$y = TX,$$

where T is a function of t only and X a function of x only. Such solutions are said to be SEPARATED into their x- and t-parts.

Since

$$\frac{\partial^2 y}{\partial t^2} = X \frac{d^2 T}{dt^2},$$

and

$$\frac{\partial^2 y}{\partial x^2} = T \frac{d^2 X}{dx^2},$$

the equation is

$$X \frac{d^2 T}{dt^2} = a^2 T \frac{d^2 X}{dx^2}$$

or

$$\frac{1}{a^2 T} \frac{d^2 T}{dt^2} = \frac{1}{X} \frac{d^2 X}{dx^2}.$$

As so displayed, the left-hand side is a function of t only and the right-hand side is a function of x only, where x, t can vary independently of each other. But it is not possible for a function of t, varying with t, to be equal to a function of x, varying with x, except when that function is a constant, varying with neither. Hence there exists a constant h such that

$$\frac{d^2 T}{dt^2} = h a^2 T,$$

$$\frac{d^2 X}{dx^2} = hX.$$

When h is *positive*, T and X appear as hyperbolic sines or cosines of t and x; when h is *negative*, they are trigonometrical sines and cosines. Having the stretched string of violin or piano in mind, we select the latter alternative, to obtain *periodic* solutions.

Writing
$$h = -n^2,$$
we get the equations
$$\frac{d^2T}{dt^2} + n^2a^2T = 0, \quad \frac{d^2X}{dx^2} + n^2X = 0,$$
with solutions
$$T = A\cos ant + B\sin ant,$$
$$X = C\cos nx + D\sin nx,$$
where A, B, C, D are constants.

[If $n = 0$, we have, exceptionally,
$$T = A + Bt, \quad X = C + Dt.]$$

The values of the constants depend on certain BOUNDARY CONDITIONS, as they are called, which limit the solutions for any particular problem. In the present example, the value of y has to vanish at the point $x = 0$ for all values of t, so that
$$C = 0.$$
Moreover, y must also vanish at the point $x = l$ for all values of t, so that
$$D\sin nl = 0.$$
Since D cannot vanish (or X, and consequently y, would be identically zero) the constant n is subject to the limitation
$$\sin nl = 0,$$
and so n assumes the form
$$n = \frac{k\pi}{l},$$
where k is some integer.

A solution of the given partial differential equation is thus
$$y = \left(A\cos\frac{ka\pi t}{l} + B\sin\frac{ka\pi t}{l}\right)\sin\frac{k\pi x}{l},$$
the arbitrary constant D having been absorbed into A and B.

A further simplification occurs if we take the motion as starting from rest at time $t = 0$; this is equivalent to the imposition of the further boundary condition
$$\frac{\partial y}{\partial t} = 0 \quad \text{when} \quad t = 0,$$

and requires the relation $B=0$.
Hence we have the solution

$$y = A\cos\frac{ka\pi t}{l}\sin\frac{k\pi x}{l}$$

for which, let us repeat, y vanishes at $x=0$ and at $x=l$ for all values of t, while $\partial y/\partial t$ vanishes at $t=0$ for all values of x. The constant A is arbitrary, and the solution satisfies all the conditions hitherto imposed so long as k is chosen to be an integer.

Having found a number of solutions of the special form XT, we proceed to derive more general solutions. It is a matter of direct substitution in the given equation

$$\frac{\partial^2 y}{\partial t^2} = a^2\frac{\partial^2 y}{\partial x^2}$$

to verify that the series

$$y = \sum_{k=1}^{N} A_k \cos\frac{k\pi at}{l}\sin\frac{k\pi x}{l}$$

is a solution for all sets of values of the arbitrary constants $A_1, A_2, \ldots, A_N$; and, further, that the three boundary conditions ($y=0$ for $x=0, l$; $\partial y/\partial t=0$ for $t=0$) remain satisfied. We have therefore obtained a solution which promises considerable generality.

It is at this point that the idea of a 'Fourier series' presents itself. The motion of the string, released, as we have supposed, from rest, depends on its initial shape at $t=0$, and the general value of y just obtained provides a solution in a case when this initial shape is given by the curve

$$y = f(x),$$

where

$$f(x) \equiv \sum_{k=1}^{N} A_k \sin\frac{k\pi x}{l}.$$

If, further, k is envisaged as taking all integral values, solutions of a still more general kind are obtained in the form of the *infinite series*

$$y = \sum_{1}^{\infty} A_k \cos\frac{k\pi at}{l}\sin\frac{k\pi x}{l}.$$

We naturally ask whether this is a form of solution that covers all cases. The string may, in theory, be held in an arbitrary shape $y=f(x)$ for a start; so, putting $t=0$ in the series, we are led to inquire whether every 'reasonable' function $f(x)$ can be expressed in the form

$$f(x)=\sum_1^\infty A_k \sin\frac{k\pi x}{l}$$

by proper choice of the constants A_k. It will appear that this is possible for a very wide variety of functions, including all the functions likely to be regarded as giving possible initial shapes for a string.

We shall call a series

$$\sum_1^\infty \left(A_k \sin\frac{k\pi x}{l}+B_k \cos\frac{k\pi x}{l}\right)$$

an INFINITE TRIGONOMETRICAL SERIES.

2. Some elementary integrals for reference. There are a few integrals whose values we shall be continually needing, and it is convenient to have a list to which reference can be made.

The numbers m, n occurring in these formulae are *integers*.

(i) *To evaluate the integral*

$$\int_{-\pi}^{\pi} \cos mx \cos nx\, dx \quad (m \neq n).$$

The integral is

$$\begin{aligned}\frac{1}{2}&\int_{-\pi}^{\pi}\{\cos(m+n)x+\cos(m-n)x\}\,dx\\ &=\frac{1}{2(m+n)}\Big[\sin(m+n)x\Big]_{-\pi}^{\pi}+\frac{1}{2(m-n)}\Big[\sin(m-n)x\Big]_{-\pi}^{\pi}\\ &=0.\end{aligned}$$

(ii) *To evaluate the integral*

$$\int_{-\pi}^{\pi}\cos^2 nx\, dx.$$

The integral is

$$\begin{aligned}\frac{1}{2}&\int_{-\pi}^{\pi}(1+\cos 2nx)\,dx\\ &=\frac{1}{2}\Big[x\Big]_{-\pi}^{\pi}+\frac{1}{4n}\Big[\sin 2nx\Big]_{-\pi}^{\pi}\\ &=\pi.\end{aligned}$$

NOTE. The particular case $n=0$ is exceptional, since

$$\int_{-\pi}^{\pi} 1\,.\,dx=2\pi.$$

(iii) *To evaluate the integral*

$$\int_{-\pi}^{\pi} \sin mx \sin nx\, dx \quad (m \neq n).$$

The integral is

$$\frac{1}{2}\int_{-\pi}^{\pi} \{\cos(m-n)x - \cos(m+n)x\}\, dx$$

$$= \frac{1}{2(m-n)}\Big[\sin(m-n)x\Big]_{-\pi}^{\pi} - \frac{1}{2(m+n)}\Big[\sin(m+n)x\Big]_{-\pi}^{\pi}$$

$$= 0.$$

(iv) *To evaluate the integral*

$$\int_{-\pi}^{\pi} \sin^2 nx\, dx.$$

The integral is $\displaystyle \frac{1}{2}\int_{-\pi}^{\pi} (1-\cos 2nx)\, dx$

$$= \frac{1}{2}\Big[x\Big]_{-\pi}^{\pi} - \frac{1}{4n}\Big[\sin 2nx\Big]_{-\pi}^{\pi}$$

$$= \pi.$$

NOTE. If $n = 0$, the value of the integral is also 0.

(v) *To evaluate the integral*

$$\int_{-\pi}^{\pi} \sin mx \cos nx\, dx \quad (m \neq n).$$

The integral is

$$\frac{1}{2}\int_{-\pi}^{\pi} \{\sin(m+n)x + \sin(m-n)x\}\, dx$$

$$= -\frac{1}{2(m+n)}\Big[\cos(m+n)x\Big]_{-\pi}^{\pi} - \frac{1}{2(m-n)}\Big[\cos(m-n)x\Big]_{-\pi}^{\pi}$$

$$= 0.$$

(vi) *To evaluate the integral*

$$\int_{-\pi}^{\pi} \sin nx \cos nx\, dx.$$

The integral is $\displaystyle \frac{1}{2}\int_{-\pi}^{\pi} \sin 2nx\, dx$

$$= -\frac{1}{4n}\Big[\cos 2nx\Big]_{-\pi}^{\pi}$$

$$= 0.$$

3. The period of a trigonometrical series. Before showing how to obtain a trigonometrical series for a *given* function in a given interval, we examine some properties of periodicity which they all possess. But first we remark that, for purposes of exposition, it is convenient to take multiples of x (rather than of $\pi x/l$ as hitherto), and so we write

$$u(x) \equiv \tfrac{1}{2}a_0 + a_1 \cos x + a_2 \cos 2x + a_3 \cos 3x + \ldots$$
$$+ b_1 \sin x + b_2 \sin 2x + b_3 \sin 3x + \ldots,$$

the coefficient $\frac{1}{2}$ being inserted before a_0 to conserve the uniformity of a formula which will appear later. We assume that the series is *convergent.*

The immediate and most striking property follows from the identities, true for all values of the integer n,

$$\cos n(x+2\pi) = \cos nx, \quad \sin n(x+2\pi) = \sin nx.$$

As a result of them, the sum of the series satisfies the *periodic relation*

$$u(x+2\pi) = u(x),$$

taking the same value for $x+2\pi$, $x+4\pi$, $x+6\pi$, ... as it does for x. In other words, *the sum is periodic in x, repeating itself at intervals of 2π.*

A diagram may help to show what is involved:

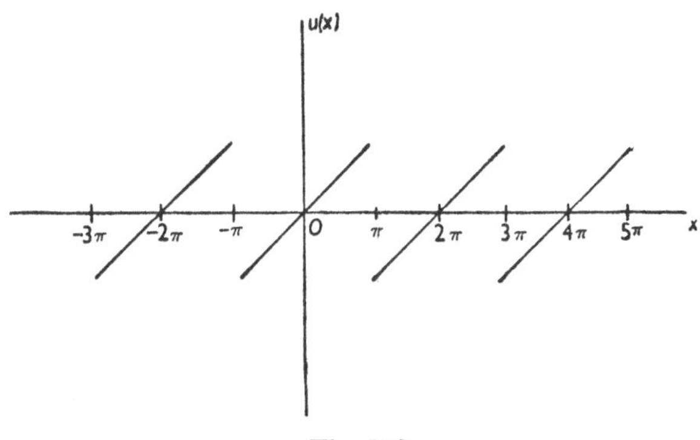

Fig. 149

Suppose that the sum is found to have the value x for the interval $-\pi < x < \pi$. The diagram (Fig. 149) gives the graph

$$u(x) = x$$

for that interval, and then shows it repeated at intervals of 2π. It appears that the series does NOT add up to the sum x for values of x in, say, the interval $\pi < x < 3\pi$, but to the sum $x - 2\pi$. Similarly, the sum of the series is $x - 4\pi$ for values of x in the interval $3\pi < x < 5\pi$, and $x + 2\pi$ for values of x in the interval $-3\pi < x < -\pi$; and so on. Putting it another way, we may say that the series

$$\tfrac{1}{2}a_0 + a_1 \cos x + a_2 \cos 2x + \ldots + b_1 \sin x + b_2 \sin 2x + \ldots$$

has the same numerical value (in magnitude and sign) for each of the values x, $x + 2\pi$, $x + 4\pi$, ...; but that numerical value, when expressed as a function of x, 'looks' different for different intervals.

It is therefore most important that, when we specify a function $f(x)$ whose trigonometrical series we require, we specify simultaneously the interval within which x must lie. The 'natural' length of such an interval, by what we have just said, is 2π.

We shall take the interval

$$-\pi < x < \pi$$

as the standard interval for exposition.

4. The calculation of the coefficients. Suppose that $f(x)$ is a given function of x defined in the interval $-\pi < x < \pi$. Suppose, too, that it *can* be expanded as a trigonometrical series in that interval in the form

$$f(x) = \tfrac{1}{2}a_0 + a_1 \cos x + a_2 \cos 2x + \ldots + a_n \cos nx + \ldots$$
$$+ b_1 \sin x + b_2 \sin 2x + \ldots + b_n \sin nx + \ldots.$$

To obtain the formulae

$$a_n = \frac{1}{\pi}\int_{-\pi}^{\pi} f(x) \cos nx\, dx \quad (n = 0, 1, 2, \ldots),$$

$$b_n = \frac{1}{\pi}\int_{-\pi}^{\pi} f(x) \sin nx\, dx \quad (n = 1, 2, \ldots).$$

The method of calculation depends basically on the formulae of § 2 (p. 189). Multiply both sides of the proposed identity by $\cos nx$ and integrate from $-\pi$ to π. On the assumption that the integral of the sum on the right-hand side is equal to the sum of the several integrals, almost all of the terms disappear (§ 2, p. 189), and we are left with the relation

$$\int_{-\pi}^{\pi} f(x) \cos nx\, dx = \pi a_n.$$

Similarly $$\int_{-\pi}^{\pi} f(x) \sin nx \, dx = \pi b_n.$$

Hence *if the expansion is possible, and if the processes of integration used in the proof are legitimate, the coefficients are given by the formulae*

$$a_n = \frac{1}{\pi} \int_{-\pi}^{\pi} f(x) \cos nx \, dx,$$

$$b_n = \frac{1}{\pi} \int_{-\pi}^{\pi} f(x) \sin nx \, dx.$$

5. The Fourier series of a function. The work up to the present does not give any information about what kinds of functions *can* be expanded in trigonometrical series of the type considered. The equations for a_n, b_n do not provide the answer, since they were calculated on the assumption that the expansion existed. They do, however, give us a starting-point for such an inquiry.

Basing our ideas on the work of § 4, we proceed as follows. Suppose that $f(x)$ is a function given in the interval $(-\pi, \pi)$. We *define* numbers a_n, b_n by the relations

$$a_n = \frac{1}{\pi} \int_{-\pi}^{\pi} f(x) \cos nx \, dx,$$

$$b_n = \frac{1}{\pi} \int_{-\pi}^{\pi} f(x) \sin nx \, dx,$$

and then form the series

$$\tfrac{1}{2}a_0 + a_1 \cos x + a_2 \cos 2x + \ldots + b_1 \sin x + b_2 \sin 2x + \ldots$$

using these coefficients. We do not now assume that the series converges to $f(x)$, but we examine it on its merits to see whether it does. This is our immediate problem.

Definition. *When $f(x)$ is a function defined in the interval $-\pi < x < \pi$, then, if*

$$a_n \equiv \frac{1}{\pi} \int_{-\pi}^{\pi} f(x) \cos nx \, dx, \quad b_n = \frac{1}{\pi} \int_{-\pi}^{\pi} f(x) \sin nx \, dx,$$

the series

$$\tfrac{1}{2}a_0 + a_1 \cos x + a_2 \cos 2x + \ldots + a_n \cos nx + \ldots$$
$$+ b_1 \sin x + b_2 \sin 2x + \ldots + b_n \sin nx + \ldots$$

is called the Fourier series *of $f(x)$ in that interval.*

We shall have to prove (i) that the Fourier series of any 'reasonable' function $f(x)$ converges for all values of x, (ii) that it converges (in general) to the value $f(x)$. For the benefit of any reader who finds the proof (given as an Appendix at the end of this chapter) rather difficult, we quote here the fundamental result:

For any given point x_0 $(-\pi < x_0 < \pi)$, *if* $\lim_{x \to x_0} f(x) \equiv f(x_0-)$ *as* x *tends to* x_0 *from values less than* x, *and if* $\lim_{x \to x_0} f(x) \equiv f(x_0+)$ *as* x *tends to* x_0 *from values greater than* x_0, *then, subject to conditions enunciated in the proof, the sum for* $x = x_0$ *of the Fourier series of* $f(x)$ *is*

$$\tfrac{1}{2}\{f(x_0-)+f(x_0+)\}.$$

If the function is continuous at x_0, then $f(x_0-) = f(x_0+)$ and the sum for $x = x_0$ is just $f(x_0)$.

The value at each of the end-points $x = -\pi$, $x = \pi$ is

$$\tfrac{1}{2}\{f(-\pi+)+f(\pi-)\};$$

see the Corollary at the end of the Appendix of this chapter (p. 211).

ILLUSTRATION 1. *To find the Fourier series of the function x in the interval $-\pi < x < \pi$.*

The coefficient a_n is given by the formula

$$\begin{aligned}\pi a_n &= \int_{-\pi}^{\pi} x \cos nx\, dx \\ &= \left[\frac{1}{n} x \sin nx\right]_{-\pi}^{\pi} - \frac{1}{n}\int_{-\pi}^{\pi} \sin nx\, dx \\ &= 0,\end{aligned}$$

as is easily verified: it should also be checked that, for the special case $n = 0$,

$$\int_{-\pi}^{\pi} x\, dx = 0.$$

Thus $$a_n = 0.$$

Also
$$\begin{aligned}\pi b_n &= \int_{-\pi}^{\pi} x \sin nx\, dx \\ &= \left[-\frac{1}{n} x \cos nx\right]_{-\pi}^{\pi} + \frac{1}{n}\int_{-\pi}^{\pi} \cos nx\, dx \\ &= -\frac{2}{n}\pi \cos n\pi \\ &= (-)^{n+1}(2\pi/n),\end{aligned}$$

so that
$$b_n = \frac{(-1)^{n+1}2}{n}.$$

Hence *the Fourier series of x is*

$$2\{\sin x - \tfrac{1}{2}\sin 2x + \tfrac{1}{3}\sin 3x - \tfrac{1}{4}\sin 4x + \ldots\}.$$

In this special case we can show, independently of the general theorem mentioned in § 5, that the series really does converge to x. Consider first the series (p. 97)

$$\log(1+z) = z - \tfrac{1}{2}z^2 + \tfrac{1}{3}z^3 - \ldots,$$

valid for $|z| \leqslant 1$, $z \neq -1$. When $z = e^{i\theta}$ $(-\pi < \theta < \pi)$, this is

$$\begin{aligned}\log(1+e^{i\theta}) &= e^{i\theta} - \tfrac{1}{2}e^{2i\theta} + \tfrac{1}{3}e^{3i\theta} - \ldots \\ &= (\cos\theta - \tfrac{1}{2}\cos 2\theta + \tfrac{1}{3}\cos 3\theta - \ldots) \\ &\qquad + i(\sin\theta - \tfrac{1}{2}\sin 2\theta + \tfrac{1}{3}\sin 3\theta - \ldots).\end{aligned}$$

Now we saw (p. 98) that

$$\log(1+re^{i\theta}) = \tfrac{1}{2}\log(1+2r\cos\theta + r^2) + i\tan^{-1}\left(\frac{r\sin\theta}{1+r\cos\theta}\right)$$

for $r \leqslant 1$, $-\pi < \theta < \pi$. Hence the imaginary part of $\log(1+e^{i\theta})$ is

$$\begin{aligned}\tan^{-1}\left(\frac{\sin\theta}{1+\cos\theta}\right) &= \tan^{-1}(\tan\tfrac{1}{2}\theta) \\ &= \tfrac{1}{2}\theta,\end{aligned}$$

for that determination of the inverse tangent which vanishes with θ; and this is the determination given by the series. It follows that

$$\sin\theta - \tfrac{1}{2}\sin 2\theta + \tfrac{1}{3}\sin 3\theta - \ldots = \tfrac{1}{2}\theta.$$

This is essentially the result just established. The ambiguity in the choice of the inverse tangent corresponds to the variations in the expression for the sum of the series for different intervals of definition.

Remark. Observe how the apparently simple series

$$\sin\theta - \tfrac{1}{2}\sin 2\theta + \tfrac{1}{3}\sin 3\theta - \ldots$$

has a discontinuous sum. The graph (p. 191) illustrates the 'break' at each point where x is an odd multiple of π. Such an example emphasizes the need for a careful examination of uniformity of convergence, such as we undertook in Chapter XXIV.

6. Odd and even functions. We saw (p. 194) that the Fourier series for the function x has zero coefficient for $\cos nx$ for all values of n. This could have been predicted, and the point is worthy of emphasis.

DEFINITION. *A function $f(x)$ is said to be* EVEN *if it satisfies the relation*

$$f(-x) \equiv f(x),$$

and ODD *if*

$$f(-x) \equiv -f(x).$$

Typical *even* functions are

$$x^2, \quad \cos x, \quad x^3 \tan x;$$

typical *odd* functions are

$$x^3, \quad \sin x, \quad \tan x.$$

Any given function $u(x)$ can be expressed as the sum of two functions $f(x)$, $g(x)$ of which $f(x)$ is even and $g(x)$ is odd. Indeed, we write

$$f(x) \equiv \tfrac{1}{2}\{u(x) + u(-x)\},$$

$$g(x) \equiv \tfrac{1}{2}\{u(x) - u(-x)\}.$$

Then

$$u(x) \equiv f(x) + g(x),$$

where

$$f(-x) \equiv f(x),$$

$$g(-x) \equiv -g(x).$$

A characteristic feature of all odd functions is embodied in the theorem:

If $f(x)$ is an odd function of x, then

$$\int_{-A}^{A} f(x)\,dx = 0$$

for any value of the constant A.

For

$$\int_{-A}^{A} f(x)\,dx = \int_{-A}^{0} f(x)\,dx + \int_{0}^{A} f(x)\,dx.$$

In the first integral on the right, write $x = -x'$ and then drop dashes. Thus

$$\int_{-A}^{0} f(x)\,dx = \int_{A}^{0} f(-x')\,(-dx')$$

$$= \int_{0}^{A} f(-x)\,dx$$

$$= -\int_{0}^{A} f(x)\,dx,$$

since $f(x)$ is an odd function. Hence

$$\int_{-A}^{A} f(x)\,dx = -\int_{0}^{A} f(x)\,dx + \int_{0}^{A} f(x)\,dx$$
$$= 0.$$

The results which follow are applications of this general property to our particular problem:

A function $f(x)$ which is EVEN *throughout the interval $-\pi < x < \pi$ has a Fourier series involving cosine terms only; and a function $f(x)$ which is* ODD *throughout the interval has a Fourier series involving sine terms only.*

If $f(x)$ is *even*, then $\qquad f(-x) \equiv f(x).$

Also, in standard notation,

$$\pi b_n \equiv \int_{-\pi}^{\pi} f(x) \sin nx\,dx$$
$$= \int_{-\pi}^{0} f(x) \sin nx\,dx + \int_{0}^{\pi} f(x) \sin nx\,dx.$$

In the first integral, write $x = -x'$ and then drop dashes; the integral is

$$\int_{\pi}^{0} f(-x) \sin(-nx)\,d(-x)$$
$$= \int_{\pi}^{0} f(-x) \sin nx\,dx$$
$$= -\int_{0}^{\pi} f(-x) \sin nx\,dx$$
$$= -\int_{0}^{\pi} f(x) \sin nx\,dx.$$

Hence $\qquad b_n = 0.$

Similarly, if $f(x)$ is *odd*

$$\pi a_n \equiv \int_{-\pi}^{\pi} f(x) \cos nx\,dx$$
$$= \int_{-\pi}^{0} f(x) \cos nx\,dx + \int_{0}^{\pi} f(x) \cos nx\,dx.$$

As before, the first integral is

$$\begin{aligned}\int_{\pi}^{0} f(-x)\cos(-nx)\,d(-x) &= -\int_{\pi}^{0} f(-x)\cos nx\,dx \\ &= \int_{0}^{\pi} f(-x)\cos nx\,dx \\ &= -\int_{0}^{\pi} f(x)\cos nx\,dx.\end{aligned}$$

Hence
$$a_n = 0.$$

7. Functions defined in the 'half-interval' $0<x<\pi$. It sometimes happens that a function $f(x)$, for which a Fourier series is required, is given only for the half-interval $0<x<\pi$. We may then 'fill up', as it were, the whole interval $-\pi<x<\pi$ by assigning to $f(x)$ values of our own choosing for the interval $-\pi<x<0$. The series so obtained will be correct for the given interval $0<x<\pi$.

In practice, one or other of two alternatives is usually adopted:

(i) The function is 'made up' to an even function by means of the definition

$$f(-x) \equiv f(x).$$

The Fourier series of $f(x)$ is then a COSINE series.

(ii) The function is 'made up' to an odd function by means of the definition

$$f(-x) \equiv -f(x).$$

The Fourier series of $f(x)$ is then a SINE series.

Thus, if we are concerned only with the interval $0<x<\pi$, we can expand any 'reasonable' function $f(x)$ in either a cosine series or a sine series, whichever we prefer.

ILLUSTRATION 2. *To find the Fourier cosine series for the function defined to have the value x in the interval $0<x<\pi$.*

The interval can be extended to $-\pi<x<\pi$ by writing

$$f(x) \equiv x \qquad (0<x<\pi),$$

$$f(x) \equiv -x \quad (-\pi<x<0).$$

Then $f(x)$ is an *even* function of x (as may also be verified graphically); so that it can be expanded in the form

$$\tfrac{1}{2}a_0 + a_1 \cos x + a_2 \cos 2x + \ldots,$$

where

$$\pi a_n = \int_{-\pi}^{\pi} f(x) \cos nx\, dx$$
$$= 2\int_0^{\pi} x \cos nx\, dx,$$

since the function is even. Thus

$$\tfrac{1}{2}\pi a_n = \left[\frac{1}{n} x \sin nx\right]_0^{\pi} - \frac{1}{n}\int_0^{\pi} \sin nx\, dx$$
$$= \frac{1}{n^2}\left[\cos nx\right]_0^{\pi}$$
$$= \frac{1}{n^2}\{\cos n\pi - 1\}$$
$$= \begin{cases} -2/n^2 & n \text{ odd,} \\ 0 & n \text{ even.} \end{cases}$$

Also

$$\pi a_0 = \int_{-\pi}^{\pi} f(x)\, dx$$
$$= 2\int_0^{\pi} x\, dx$$
$$= \pi^2.$$

Hence, for the interval $0<x<\pi$, the expansion of x as a *cosine* series is

$$\tfrac{1}{2}\pi - \frac{4}{\pi}\left\{\cos x + \frac{1}{3^2}\cos 3x + \ldots + \frac{1}{(2n+1)^2}\cos(2n+1)x + \ldots\right\}.$$

ILLUSTRATION 3. *To find the Fourier series of the function $f(x)$ such that*

$$f(x) = 0 \quad (-\pi < x \leqslant 0),$$
$$f(x) = 1 \quad (0 < x \leqslant \pi),$$

and to deduce the formula

$$1 - \tfrac{1}{3} + \tfrac{1}{5} - \tfrac{1}{7} + \ldots = \tfrac{1}{4}\pi.$$

The series is

$$\tfrac{1}{2}a_0 + a_1 \cos x + a_2 \cos 2x + \ldots + b_1 \sin x + b_2 \sin 2x + \ldots,$$

where
$$\pi a_n = \int_{-\pi}^{\pi} f(x) \cos nx\,dx,$$
$$\pi b_n = \int_{-\pi}^{\pi} f(x) \sin nx\,dx.$$

Thus
$$\pi a_n = \int_{-\pi}^{0} 0 . \cos nx\,dx + \int_{0}^{\pi} 1 . \cos nx\,dx \quad (n \neq 0)$$
$$= 0;$$
$$\pi a_0 = \int_{-\pi}^{0} 0 . dx + \int_{0}^{\pi} 1 . dx = \pi,$$

so that
$$a_0 = 1.$$

Also
$$\pi b_n = \int_{-\pi}^{0} 0 . \sin nx\,dx + \int_{0}^{\pi} 1 . \sin nx\,dx$$
$$= -\frac{1}{n}\Big[\cos nx\Big]_0^{\pi}$$
$$= \frac{1}{n}(1 - \cos n\pi).$$

Hence the series is
$$\tfrac{1}{2} + \frac{1}{\pi}\{2 \sin x + \tfrac{2}{3} \sin 3x + \tfrac{2}{5} \sin 5x + \ldots\}$$
$$= \tfrac{1}{2} + \frac{2}{\pi}\{\sin x + \tfrac{1}{3} \sin 3x + \tfrac{1}{5} \sin 5x + \ldots\}.$$

In particular, when $x = \frac{1}{2}\pi$, we have the relation
$$1 = \tfrac{1}{2} + \frac{2}{\pi}(1 - \tfrac{1}{3} + \tfrac{1}{5} - \ldots),$$

so that
$$1 - \tfrac{1}{3} + \tfrac{1}{5} - \tfrac{1}{7} + \ldots = \tfrac{1}{4}\pi.$$

Note, incidentally, that, at the point $x = 0$, the value of $f(0-)$ is 0 and the value of $f(0+)$ is 1. Also the sum of the Fourier series for $x = 0$ is $\frac{1}{2}$, thus verifying the formula $\frac{1}{2}\{f(0-) + f(0+)\}$.

ILLUSTRATION 4. *To find the Fourier series of the function $f(x)$ such that*
$$f(x) = 1 \quad (-\pi < x \leqslant 0),$$
$$f(x) = x \quad (0 < x \leqslant \pi),$$

and to deduce the formula
$$1 + \frac{1}{3^2} + \frac{1}{5^2} + \frac{1}{7^2} + \ldots = \tfrac{1}{8}\pi^2.$$

The series is

$$\tfrac{1}{2}a_0 + a_1 \cos x + a_2 \cos 2x + \ldots + b_1 \sin x + b_2 \sin 2x + \ldots,$$

where

$$\pi a_n = \int_{-\pi}^{\pi} f(x) \cos nx \, dx,$$

$$\pi b_n = \int_{-\pi}^{\pi} f(x) \sin nx \, dx.$$

Thus

$$\pi a_n = \int_{-\pi}^{0} \cos nx \, dx + \int_{0}^{\pi} x \cos nx \, dx \quad (n \neq 0)$$

$$= \left[\frac{1}{n} \sin nx\right]_{-\pi}^{0} + \left[\frac{1}{n^2} \cos nx + \frac{x}{n} \sin nx\right]_{0}^{\pi}$$

$$= \frac{1}{n^2} \cos n\pi - \frac{1}{n^2} = -\frac{1}{n^2}(1 - \cos n\pi);$$

$$\pi a_0 = \int_{-\pi}^{0} dx + \int_{0}^{\pi} x \, dx = \pi + \tfrac{1}{2}\pi^2.$$

Also

$$\pi b_n = \int_{-\pi}^{0} \sin nx \, dx + \int_{0}^{\pi} x \sin nx \, dx$$

$$= \left[-\frac{1}{n} \cos nx\right]_{-\pi}^{0} + \left[\frac{1}{n^2} \sin nx - \frac{x}{n} \cos nx\right]_{0}^{\pi}$$

$$= -\frac{1}{n}(1 - \cos n\pi) - \frac{\pi}{n} \cos n\pi.$$

Hence the series is (after a little reduction)

$$\tfrac{1}{2} + \tfrac{1}{4}\pi - \frac{2}{\pi}\left(\cos x + \frac{1}{3^2} \cos 3x + \frac{1}{5^2} \cos 5x + \ldots\right)$$
$$+ \left\{\frac{(\pi - 2)}{\pi} \sin x - \tfrac{1}{2} \sin 2x + \frac{(\pi - 2)}{3\pi} \sin 3x - \tfrac{1}{4} \sin 4x + \frac{(\pi - 2)}{5\pi} \sin 5x - \ldots\right\}.$$

In particular, the series when $x = 0$ is

$$\tfrac{1}{2} + \tfrac{1}{4}\pi - \frac{2}{\pi}\left(1 + \frac{1}{3^2} + \frac{1}{5^2} + \ldots\right),$$

and $f(0-) = 1, f(0+) = 0$, so that the sum is $\tfrac{1}{2}$. Hence

$$\tfrac{1}{2} = \tfrac{1}{2} + \tfrac{1}{4}\pi - \frac{2}{\pi}\left(1 + \frac{1}{3^2} + \frac{1}{5^2} + \ldots\right),$$

so that

$$1 + \frac{1}{3^2} + \frac{1}{5^2} + \ldots, = \tfrac{1}{8}\pi^2.$$

APPENDIX TO CHAPTER XXVIII

THE CONVERGENCE TO $f(x)$ OF ITS FOURIER SERIES

1. The form of $f(x)$. The type of function with which we shall be concerned is illustrated in the diagram (Fig. 150). The function $f(x)$ is bounded in the interval $-\pi < x < \pi$ and is continuous* save possibly at a finite number of points; its differential coefficient

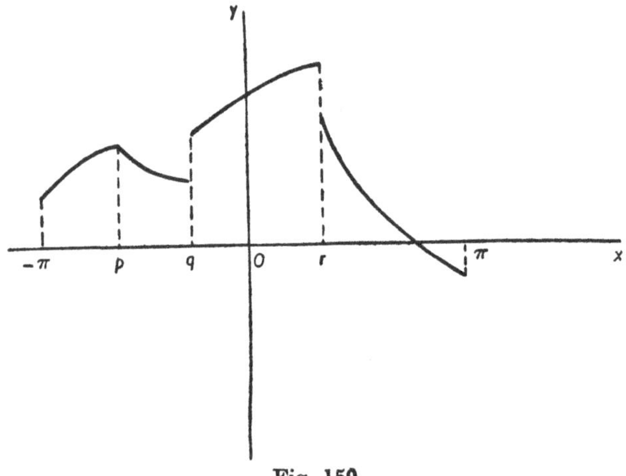

Fig. 150

is also bounded and continuous save possibly at a finite number of points. The interval $-\pi < x < \pi$ may therefore be divided into a finite number of sub-intervals

$$-\pi < x < p, \quad p < x < q, \quad q < x < r, \quad \ldots, \quad w < x < \pi,$$

in each of which $f(x)$ and its differential coefficient $f'(x)$ are both continuous.

Consider a typical interval $p < x < q$. We assume that, as $x \to p$ in this interval (that is, through values greater than p), the function is of such a type that the limit

$$\lim_{x \to p} f(x) \quad (x > p)$$

* For 'bounded and continuous', compare the remark on p. 138.

of the function exists; similarly we assume that, as $x \to q$ in the interval (that is through values less than q), the limit

$$\lim_{x \to q} f(x) \quad (x < q)$$

exists.

We now make the agreement that, *for the purposes of a calculation in the interval* $p < x < q$, the function at each end-point is to be given the value to which it tends from within that interval. (Informally: the arcs in the diagram are to be given *for each interval* that value at the end-point which would be expected from continuity.) Thus we assume that the limits

$$\lim_{x \to p} f(x) \quad (x > p),$$

$$\lim_{x \to q} f(x) \quad (x < q)$$

exist, and we denote them by the values

$$f(p+), \quad f(q-)$$

respectively.

This process may involve assigning one value $f(q-)$ to $f(x)$ at q for the interval (p, q) and a *different* value $f(q+)$ for the interval (q, r). The actual value $f(q)$ may be different from either; but fortunately it turns out that the value of $f(x)$ at a single point like q does not affect the behaviour of the Fourier series, so that the conventions just described can be adopted with safety.

For the differential coefficients at p, q we shall use a natural extension of ordinary ideas. We define the RIGHT-HAND DIFFERENTIAL COEFFICIENT at, say, p to be the limit, if it exists,

$$\lim_{h \to 0} \frac{f(p+h) - f(p+)}{h},$$

as h tends to zero through *positive* values; and the LEFT-HAND DIFFERENTIAL COEFFICIENT at, say, q to be the limit, if it exists,

$$\lim_{h \to 0} \frac{f(q+h) - f(q-)}{h},$$

as h tends to zero through *negative* values. When we talk about differential coefficients at points such as $p, q, r, \ldots$, it is these limits that are meant.

We agree, then, that, with these conventions, the function $f(x)$ and its differential coefficient are uniquely defined, bounded, and continuous throughout the whole of each interval

$$p \leqslant x \leqslant q.$$

2. Three lemmas. Before coming to the main theorem, we prove three lemmas which we shall require on the way.

(i) *The value of the integral*

$$\int_a^b g(\theta)\, d\theta$$

of a bounded function $g(\theta)$ is not affected if the value of $g(\theta)$ is altered at a finite number of points.

Suppose that $g(\theta)$ is altered at the k points $\theta_1, \theta_2, \ldots, \theta_k$, and denote the new function by $h(\theta)$. Suppose, too, that

$$|g(\theta)| \leqslant M, \quad |h(\theta)| \leqslant M$$

throughout the interval $a \leqslant \theta \leqslant b$. Surround each of the points $\theta_1, \theta_2, \ldots, \theta_k$ by an interval of length δ, say. The contribution from these intervals to the integral of $g(\theta)$ is numerically less than $kM\delta$, and this can be made less than a given positive number ϵ by choosing δ so that

$$\delta < (\epsilon/kM).$$

The contribution from these intervals to the integral of $h(\theta)$ is then also less than ϵ. Hence the difference between the integrals is certainly less than 2ϵ; and since ϵ may be chosen as small as we please, the two integrals are equal in value.

This lemma will usually be incorporated into subsequent arguments without explicit mention.

(ii) *If the bounded function $g(\theta)$ has a differential coefficient which is bounded and continuous throughout the interval $a \leqslant x \leqslant b$, then*

$$\lim_{k\to\infty} \int_a^b g(\theta) \sin k\theta\, d\theta = 0, \quad \lim_{k\to\infty} \int_a^b g(\theta) \cos k\theta\, d\theta = 0.$$

On integration by parts we have, for the first integral,

$$\int_a^b g(\theta) \sin k\theta\, d\theta = \frac{1}{k}\Big[-g(\theta)\cos k\theta\Big]_a^b + \frac{1}{k}\int_a^b g'(\theta) \cos k\theta\, d\theta,$$

so that

$$\left|\int_a^b g(\theta) \sin k\theta\, d\theta\right| \leqslant \frac{1}{k}|g(b)\cos kb| + \frac{1}{k}|g(a)\cos ka| + \frac{1}{k}(b-a)N,$$

where N is the greatest value of $|g'(\theta)|$ in the interval. Hence

$$\lim_{k\to\infty}\int_a^b g(\theta)\sin k\theta\,d\theta=0.$$

Similarly
$$\lim_{k\to\infty}\int_a^b g(\theta)\cos k\theta\,d\theta=0.$$

COROLLARY. *If α is a constant, then*

$$\lim_{k\to\infty}\int_a^b g(\theta)\sin k(\theta-\alpha)\,d\theta=0.$$

This follows at once from the formula

$$\sin k(\theta-\alpha)=\cos k\alpha\,\sin k\theta-\sin k\alpha\,\cos k\theta.$$

(iii) *If $0<x<2\pi$, then the sum of the infinite series*

$$\sin x+\tfrac{1}{2}\sin 2x+\ldots+\frac{1}{n}\sin nx+\ldots$$

is $\tfrac{1}{2}(\pi-x)$. (The formula is not true for $x=0$ or for $x=2\pi$.)

The result can be derived from the expansion for $\log(1-z)$; the proof which follows* is from elementary principles.

The sum of the first n terms of the series is

$$U_n(x)\equiv\sin x+\tfrac{1}{2}\sin 2x+\ldots+\frac{1}{n}\sin nx$$
$$=\int_0^x(\cos\theta+\cos 2\theta+\ldots+\cos n\theta)\,d\theta.$$

Multiply the cosine series by $2\sin\tfrac{1}{2}\theta$, use the formula

$$2\sin\tfrac{1}{2}\theta\cos\lambda\theta=\sin(\lambda+\tfrac{1}{2})\,\theta-\sin(\lambda-\tfrac{1}{2})\,\theta$$

and sum:
$$U_n(x)=\int_0^x\frac{-\sin\tfrac{1}{2}\theta+\sin(n+\tfrac{1}{2})\,\theta}{2\sin\tfrac{1}{2}\theta}\,d\theta$$
$$=-\tfrac{1}{2}x+\int_0^x\frac{\sin(n+\tfrac{1}{2})\,\theta}{2\sin\tfrac{1}{2}\theta}\,d\theta.$$

Moreover, from the series itself,

$$U_n(\pi)=0,$$

so that, substituting $x=\pi$ in the expression for $U_n(x)$,

$$0=-\tfrac{1}{2}\pi+\int_0^\pi\frac{\sin(n+\tfrac{1}{2})\,\theta}{2\sin\tfrac{1}{2}\theta}\,d\theta.$$

* I am indebted for it to Dr J. C. Burkill.

Subtract this equation from the formula for $U_n(x)$:

$$U_n(x) = \tfrac{1}{2}(\pi - x) - \int_x^\pi \frac{\sin(n+\frac{1}{2})\theta}{2\sin\frac{1}{2}\theta}\, d\theta.$$

But $1/\sin\frac{1}{2}\theta$ is bounded and has a bounded and continuous differential coefficient in the interval of integration (where $0 < x < 2\pi$), and so, by the second Lemma (p. 204),

$$\lim_{n\to\infty} \int_x^\pi \frac{\sin(n+\frac{1}{2})\theta}{2\sin\frac{1}{2}\theta}\, d\theta = 0.$$

Hence
$$U(x) \equiv \lim_{n\to\infty} U_n(x)$$
$$= \tfrac{1}{2}(\pi - x).$$

3. The convergence of the Fourier series for $f(x)$. Let $f(x)$ be a function subject to the conditions enumerated in § 1 (p. 202). Its Fourier series is

$$\tfrac{1}{2}a_0 + a_1 \cos x + a_2 \cos 2x + \dots + a_n \cos nx + \dots$$
$$+ b_1 \sin x + b_2 \sin 2x + \dots + b_n \sin nx + \dots,$$

where $$a_n = \frac{1}{\pi}\int_{-\pi}^{\pi} f(\theta)\cos n\theta\, d\theta, \quad b_n = \frac{1}{\pi}\int_{-\pi}^{\pi} f(\theta)\sin n\theta\, d\theta.$$

We prove that *the Fourier series of $f(x)$ converges and its sum is*

$$\tfrac{1}{2}\{f(x+) + f(x-)\}.$$

Denote by S_n the sum of the terms up to and including those in a_n and b_n. Then

$$\pi S_n = \int_{-\pi}^{\pi} f(\theta)\{\tfrac{1}{2} + \cos\theta\cos x + \cos 2\theta \cos 2x + \dots$$
$$+ \sin\theta \sin x + \sin 2\theta \sin 2x + \dots\}\, d\theta$$
$$= \int_{-\pi}^{\pi} f(\theta)\{\tfrac{1}{2} + \cos(\theta - x) + \cos 2(\theta - x) + \dots + \cos n(\theta - x)\}\, d\theta$$
$$= \int_{-\pi}^{\pi} f(\theta)\left\{\frac{\sin(n+\frac{1}{2})(\theta - x)}{2\sin\frac{1}{2}(\theta - x)}\right\} d\theta.$$

(Compare the summation in § 2, p. 205.)

Let δ, δ' be two positive numbers chosen sufficiently small to ensure that (whether $f(\theta)$ is continuous or discontinuous at the point x under consideration) there is no discontinuity of $f(\theta)$ or of $f'(\theta)$ in either of the two intervals

$$x-\delta' \leqslant \theta < x, \quad x < \theta \leqslant x+\delta.$$

Divide the interval of integration into four parts:

$$\int_{-\pi}^{\pi} \equiv \int_{-\pi}^{x-\delta'} + \int_{x-\delta'}^{x} + \int_{x}^{x+\delta} + \int_{x+\delta}^{\pi}.$$

It will be established that the pith of the integral lies in the two parts $\int_{x-\delta'}^{x}, \int_{x}^{x+\delta}$, the contributions from the two 'outer' parts being negligible. We must now make this statement more precise.

Consider the integral

$$\int_{x}^{x+\delta} f(\theta) \left\{ \frac{\sin (n+\frac{1}{2})(\theta - x)}{2 \sin \frac{1}{2}(\theta - x)} \right\} d\theta.$$

Write $\theta - x = u$. The integral is

$$\int_{0}^{\delta} f(x+u) \frac{\sin (n+\frac{1}{2}) u}{2 \sin \frac{1}{2} u} du.$$

We have so far placed one restriction on δ. This ensures that the function $f(\theta)$ has a differential coefficient throughout the interval $x < \theta < x+\delta$, as well as being itself continuous throughout $x \leqslant \theta \leqslant x+\delta$. The conditions of the mean-value theorem* are therefore satisfied, and so there exists a number ξ, where $x < \xi < x+\delta$, such that

$$f(x+u) = f(x+) + uf'(\xi),$$

where $f(x+)$ is the value to which $f(\theta)$ tends as θ tends to x through positive values.

We are therefore led to the sum

$$f(x+) \int_{0}^{\delta} \frac{\sin (n+\frac{1}{2}) u}{2 \sin \frac{1}{2} u} du + \int_{0}^{\delta} uf'(\xi) \frac{\sin (n+\frac{1}{2}) u}{2 \sin \frac{1}{2} u} du,$$

and we begin by examining the numerical value of the second of these integrals, restricting δ further to make that value negligibly small.

* See Volume I, p. 61. A more precise statement may be obtained in books giving detailed analytical treatments; for example, W. L. Ferrar, *Differential Calculus* (1956), p. 94.

Now
$$\left|\int_0^\delta uf'(\xi)\frac{\sin(n+\frac{1}{2})u}{2\sin\frac{1}{2}u}\,du\right|$$
$$\leqslant\left|\max.\frac{\frac{1}{2}u}{\sin\frac{1}{2}u}\right|.\left|\max.f'(\xi)\right|.\left|\max.\sin(n+\tfrac{1}{2})u\right|.\delta;$$

and we have at once the inequalities

$$|\max.f'(\xi)|<K,$$

say, since $f'(\theta)$ is bounded, and

$$|\max.\sin(n+\tfrac{1}{2})u|\leqslant 1.$$

For the first maximum, we recall the inequality (Volume I, p. 53)

$$\sin\phi>\phi-\tfrac{1}{6}\phi^3\quad(\phi>0),$$

so that (with $1-\frac{1}{6}\phi^2$ taken to be positive)

$$\frac{\phi}{\sin\phi}<\frac{1}{1-\frac{1}{6}\phi^2},$$

or, here,
$$\frac{\frac{1}{2}u}{\sin\frac{1}{2}u}<\frac{1}{1-\frac{1}{24}u^2}.$$

The right-hand side can be kept bounded; it is, for example, less than 2 if
$$1-\tfrac{1}{24}u^2>\tfrac{1}{2}$$
or*
$$u^2<12.$$

In the nature of the case, u is less than δ, the upper limit of integration, and so we can certainly make

$$\left|\max.\frac{\frac{1}{2}u}{\sin\frac{1}{2}u}\right|<2$$

by taking
$$\delta<\sqrt{12},$$

which is much larger than we shall ultimately want. Thus, so long as $\delta<\sqrt{12}$, we have the inequality

$$\left|\int_0^\delta uf'(\xi)\frac{\sin(n+\frac{1}{2})u}{2\sin\frac{1}{2}u}\,du\right|<2K\delta.$$

Suppose now that ϵ is a given (small) positive number. We can ensure that this numerical value is less than ϵ by taking

$$2K\delta<\epsilon,$$

* The value 12 is much higher than is necessary, but it arises naturally out of the work, and is good enough.

or
$$\delta < (\epsilon/2K).$$

This is a more stringent condition than we have yet imposed on δ, so all earlier consequences remain valid.

Having selected a *definite* value of δ, subject to this condition, we turn next to the integral

$$\int_0^\delta \frac{\sin (n+\frac{1}{2})\, u}{2 \sin \frac{1}{2}u}\, du.$$

We proved (pp. 205–6) that, if (with a change of notation)

$$\int_0^\delta \frac{\sin (n+\frac{1}{2})\, u}{2 \sin \frac{1}{2}u}\, du = U_n(\delta) + \tfrac{1}{2}\delta$$

for given δ, then, as $n \to \infty$,

$$U_n(\delta) \to \tfrac{1}{2}(\pi - \delta).$$

Hence
$$\int_0^\delta \frac{\sin (n+\frac{1}{2})\, u}{2 \sin \frac{1}{2}u}\, du \to \tfrac{1}{2}\pi.$$

Finally, let us return to the 'outside' integral (p. 207)

$$\int_{x+\delta}^{\pi} f(\theta) \left\{ \frac{\sin (n+\frac{1}{2})\, (\theta - x)}{2 \sin \frac{1}{2}(\theta - x)} \right\} d\theta.$$

The interval $x+\delta \leqslant \theta \leqslant \pi$ may be divided into a number of subintervals of the type described in § 1 (p. 202), say

$$x+\delta \leqslant \theta \leqslant q, \quad q \leqslant \theta \leqslant r, \quad \ldots, \quad w \leqslant \theta \leqslant \pi,$$

and we express the integral over $(x+\delta, \pi)$ as the sum of the integrals over these subintervals. In dealing with each one, we can suppose that the values of $f(\theta)$ and its differential coefficient $f'(\theta)$ at the end-points are defined so as to be continuous throughout it, as described on p. 203; in virtue of Lemma 1, any changes required for this purpose do not affect the values of the integrals.

A typical integral is

$$\int_q^r f(\theta) \left\{ \frac{\sin (n+\frac{1}{2})\, (\theta - x)}{2 \sin \frac{1}{2}(\theta - x)} \right\} d\theta,$$

or
$$\int_q^r g(\theta) \sin \{(n+\tfrac{1}{2})\, (\theta - x)\}\, d\theta,$$

where
$$g(\theta) \equiv \frac{f(\theta)}{2 \sin \frac{1}{2}(\theta - x)}.$$

Now $g(\theta)$ and its differential coefficient with respect to θ are bounded and continuous in each of the intervals

$$x+\delta \leqslant \theta \leqslant q, \quad q \leqslant \theta \leqslant r, \quad \ldots, \quad w \leqslant \theta \leqslant \pi,$$

in virtue of the conditions satisfied by $f(\theta)$, $f'(\theta)$ and the fact that $\sin \frac{1}{2}(\theta-x)$ does not vanish anywhere in the interval $x+\delta \leqslant \theta \leqslant \pi$. Hence, by the corollary to Lemma (ii) of § 2 (p. 204),

$$\lim_{n\to\infty} \int_q^r g(\theta) \sin\{(n+\tfrac{1}{2})(\theta-x)\}\, d\theta = 0,$$

and so, adding for all such integrals,

$$\lim_{n\to\infty} \int_{x+\delta}^{\pi} f(\theta) \left\{\frac{\sin (n+\frac{1}{2})(\theta-x)}{2\sin\frac{1}{2}(\theta-x)}\right\} d\theta = 0.$$

We are now in a position to give a résumé and completion of the whole argument. We have expressed πS_n (p. 207) as the sum of four integrals, briefly written

$$\int_{-\pi}^{x-\delta'} + \int_{x-\delta'}^{x} + \int_{x}^{x+\delta} + \int_{x+\delta}^{\pi}.$$

We have (p. 208) shown how to choose δ so that the third integral is the sum of two parts, of which one is less than ϵ for all n. When this has been done, the value of n is allowed to increase; the second part of this integral then tends to $\frac{1}{2}\pi f(x+)$, and the fourth integral tends to zero. Hence, when n is large enough, the sum of the third and fourth integrals will differ from $\frac{1}{2}\pi f(x+)$ by less than 2ϵ. Similarly the sum of the first and second can be made to differ from $\frac{1}{2}\pi f(x-)$ by less than 2ϵ. Hence, since ϵ may be as small as we please, we have the fundamental theorem:

The sum of the Fourier series is given by the expression

$$S = \tfrac{1}{2}\{f(x+)+f(x-)\}.$$

In particular, *the sum is $f(x)$ at a point where the function is continuous.*

Note that, whatever value we may have assigned to $f(x)$ at the points $p, q, r, \ldots,$ the sum of the Fourier series at those points is $\frac{1}{2}\{f(p+)+f(p-)\}$ and so on. As we saw (Lemma (i) of § 2), the values assigned to the function at such isolated points do not affect the calculations of the integrals on which the formula depends.

COROLLARY. Since the Fourier series of $f(x)$ is periodic, with period 2π, it is easy to show in a similar way that *the sum of the series at each of the end-points* $x=-\pi$, $x=+\pi$ *of the interval* $(-\pi,\pi)$ *is*

$$\tfrac{1}{2}\{f(-\pi+)+f(\pi-)\}.$$

REVISION EXAMPLES XXIII

1. Find a Fourier series of period 2π which represents the function $\sin tx$ in the interval $(-\pi,\pi)$ when t is not an integer.

What is the sum of the series when $x=\pm\pi$?

What function does the Fourier series represent in the range $(\pi, 3\pi)$?

2. Find the Fourier series of period 2π which gives a function equal to 0 for $-\pi<x<0$ and equal to $\cosh x$ for $0<x<\pi$.

By considering the values of this series at the points of discontinuity, deduce that

$$\sum_1^\infty \frac{(-1)^n}{1+n^2}=\frac{1}{2}\left(\frac{\pi}{\sinh\pi}-1\right),\quad \sum_1^\infty \frac{1}{1+n^2}=\frac{1}{2}\left(\frac{\pi}{\tanh\pi}-1\right).$$

3. Find a Fourier sine series for the function $f(x)$ defined by the relations

$$f(x)=2x/\pi \qquad (0\leqslant x\leqslant \tfrac{1}{2}\pi),$$
$$f(x)=2-(2x/\pi) \quad (\tfrac{1}{2}\pi\leqslant x\leqslant \pi).$$

What function does the series represent in the interval

$$2\pi\leqslant x\leqslant 3\pi?$$

4. Show that the Fourier expansion of $|x|$ in the interval $-\pi<x<\pi$ is

$$\frac{\pi}{2}-\frac{4}{\pi}\left(\cos x+\frac{\cos 3x}{3^2}+\frac{\cos 5x}{5^2}+\dots\right).$$

State what function is represented by this expansion in the interval $(n-2)\pi<x<n\pi$, where n is an integer. Use the expansion of $|x|$ in $-\pi<x<\pi$ to obtain the following series for π^2:

$$\frac{\pi^2}{8}=1+\frac{1}{3^2}+\frac{1}{5^2}+\dots.$$

Hence or otherwise show that the Fourier expansion for $|x|$ in $-\pi<x<\pi$ can be extended to the closed interval $-\pi\leqslant x\leqslant\pi$. What special property of $|x|$ enables this extension to be made?

5. Express $x \sin x$ as a Fourier series of the form

$$\tfrac{1}{2}a_0 + \sum_{n=1}^{\infty} (a_n \cos nx + b_n \sin nx)$$

in the range $-\pi \leqslant x \leqslant \pi$.

What is the function represented by the series in the range $\pi \leqslant x \leqslant 3\pi$?

By taking a particular value of x, deduce that

$$\frac{1}{1.3} - \frac{1}{3.5} + \frac{1}{5\ 7} - \dots = \tfrac{1}{4}(\pi - 2).$$

6. Express x^2 as a Fourier sine series of the form

$$b_1 \sin x + b_2 \sin 2x + b_3 \sin 3x + \dots$$

in the range $0 \leqslant x \leqslant \pi$. Does the value of the series agree with that of the function at the end-points?

What is the function represented by the series in the range $\pi < x < 2\pi$?

By considering the value of the series at $x = \tfrac{1}{2}\pi$, deduce that

$$1 - \frac{1}{3^3} + \frac{1}{5^3} - \dots = \frac{\pi^3}{32}.$$

(It may be assumed that $\tfrac{1}{4}\pi = 1 - \tfrac{1}{3} + \tfrac{1}{5} - \dots$.)

7. Obtain a series of the form

$$\tfrac{1}{2}a_0 + \sum_{n=1}^{\infty} a_n \cos nx$$

whose sum is x^2 when $-\pi < x < \pi$.

Sketch the graph of the function represented by the series in the range $(-3\pi, 3\pi)$.

Deduce from your work that

$$1 - \frac{1}{2^2} + \frac{1}{3^2} - \frac{1}{4^2} + \dots = \frac{\pi^2}{12}.$$

8. Express the function $\pi - 2x$, in the range $0 < x < \pi$, (i) as a series of cosines of $x, 2x, 3x, \dots$; (ii) as a series of sines and cosines of $2x, 4x, 6x, \dots$.

State (without proof but with an indication how you obtain the result) what will be the sums of the two series you obtain (*a*) at $x = 0$, (*b*) in the range $-\pi < x < 0$.

9. Obtain a series of the form

$$\tfrac{1}{2}a_0 + \sum_{n=1}^{\infty} a_n \cos nx,$$

whose sum is 1 when $0 < x < \frac{2}{3}\pi$, and 0 when $\frac{2}{3}\pi < x < \pi$. State, with reasons, the sum of the series when $x = 0$ and when $x = \pi$.

Obtain the sum of the series

$$1 - \tfrac{1}{2} + \tfrac{1}{4} - \tfrac{1}{5} + \tfrac{1}{7} - \tfrac{1}{8} + \tfrac{1}{10} - \tfrac{1}{11} + \dots.$$

What is the sum of the Fourier series when $x = \frac{2}{3}\pi$? Sketch the graph of the function represented by the series for all real values of x.

10. Show that the function $a^2 - x^2$ can be represented in the range $-a \leqslant x \leqslant a$ by the trigonometrical series

$$4a^2\left\{\frac{1}{6} - \frac{1}{\pi^2}\sum_1^{\infty}\frac{(-1)^n}{n^2}\cos\frac{n\pi x}{a}\right\}.$$

11. A square has two opposite vertices at the points $(\pm\pi, 0)$. Find a Fourier cosine series which represents the ordinate of any point of the perimeter on the positive side of the axis of x.

12. Calculate the coefficients c_n in the series

$$\sum_{n=0}^{\infty} c_n \cos\frac{n\pi x}{a}$$

that represents the function $x(a-x)$ in the range $0 \leqslant x \leqslant a$.

Deduce the value of $\sum_{n=1}^{\infty}\frac{1}{n^2}$.

13. Prove that, if k is not an integer, the expansion of $\cos kx$ in series of cosines of multiples of x in the range $(0, \pi)$ of x is

$$\frac{2k\sin k\pi}{\pi}\left\{\frac{1}{2k^2} - \frac{\cos x}{k^2 - 1^2} + \frac{\cos 2x}{k^2 - 2^2} - \dots\right\}.$$

By sketching a graph or otherwise show what function is represented by the series in the ranges $(-\pi, 0)$ and $(\pi, 2\pi)$ of x.

14. Given that $0 < \alpha < \pi$, find a Fourier sine series for the function $f(x)$ defined by the conditions

(i) $f(x) = -x$ if $0 \leqslant x < \alpha$,

(ii) $f(x) = \frac{1}{2}\pi - \alpha$ if $x = \alpha$,

(iii) $f(x) = \pi - x$ if $\alpha < x \leqslant \pi$.

Illustrate by a rough graph the function which the series represents in $-\pi \leqslant x \leqslant 3\pi$.

15. Find the Fourier series of the function which has period 2π and is equal to x^3 when $-\pi \leqslant x < \pi$.

What function does the series represent when

$$(2k-1)\pi \leqslant x < (2k+1)\pi?$$

16. Find the Fourier series of the function which has period 2π and is equal to $\pi^2 - x^2$ when $-\pi \leqslant x < \pi$.

Deduce the sum of the series

$$1 - \frac{1}{2^2} + \frac{1}{3^2} - \frac{1}{4^2} + \dots.$$

17. Expand the function x^2 in the interval $-\pi \leqslant x \leqslant \pi$ as the sum of a series of cosines of multiples of x.

Prove that the locus given by

$$\sum_{n=1}^{\infty} \frac{(-1)^n}{n^2} \sin nx \sin ny = 0$$

is two systems of lines at right angles dividing the (x, y) plane into squares of area π^2.

18. Prove that, for all real values of x,

$$|\sin x| = \frac{2}{\pi} - \frac{4}{\pi} \sum_{n=1}^{\infty} \frac{\cos 2nx}{4n^2-1}.$$

19. Prove that, if

$$f(x) = \frac{\sinh \lambda x}{\cosh \lambda a} \qquad (0 \leqslant x \leqslant a),$$

$$f(x) = \frac{\sinh \lambda(2a-x)}{\cosh \lambda a} \qquad (a \leqslant x \leqslant 2a),$$

then

$$f(x) = 8a\lambda \sum_{n=1}^{\infty} \frac{\sin\left(\frac{n\pi}{2}\right) \sin\left(\frac{n\pi x}{2a}\right)}{4a^2\lambda^2 + n^2\pi^2}.$$

20. The spot of a cathode-ray tube moves uniformly from A to B through a distance l along a straight line and then flicks instantaneously back to A. If this cycle takes a time T and is repeated continuously and indefinitely for all time, find the Fourier analysis of the function that represents the dependence upon time of the displacement of the spot from A.

If all harmonics of frequency greater than $4/T$ are suppressed, sketch the resulting dependence of displacement upon time.

21. A point moves in a straight line with initial velocity u. At the end of each second during the motion its velocity receives a sudden increment u. Prove that the velocity at any time t after the motion has begun is

$$\tfrac{1}{2}u + ut + \frac{u}{\pi}\sum_{n=1}^{\infty} \frac{\sin 2n\pi t}{n}.$$

22. The function $y(x)$ is defined in the range $-1 \leqslant x \leqslant 1$ by

$$y = \tfrac{1}{2}\epsilon^{-1} \quad (-\epsilon \leqslant x \leqslant \epsilon),$$

$$y = 0 \quad (-1 \leqslant x < -\epsilon \text{ and } \epsilon < x \leqslant 1).$$

Show that the Fourier expansion of y in the range $-1 \leqslant x \leqslant 1$ is given by

$$y = \frac{1}{2} + \sum_{n=1}^{\infty} \frac{\sin(n\pi\epsilon)}{n\pi\epsilon} \cos(n\pi x).$$

The function $z(x)$ is defined in the range $-1 \leqslant x \leqslant 1$ by

$$z = \int_0^x y(x)\,dx.$$

Draw a graph of $z(x)$ and find its Fourier expansion in $-1 < x < 1$. Discuss the behaviour of $z(x)$ and of its Fourier expansion as $\epsilon \to 0$.

SECTION 3

LAPLACE'S EQUATION AND RELATED EQUATIONS

CHAPTER XXIX

THE TRANSFORMATION OF LAPLACE'S EQUATION

1. The equations to be investigated. We use the notation

$$\nabla^2 U$$

to denote the function
$$\frac{\partial^2 U}{\partial x^2},$$

or
$$\frac{\partial^2 U}{\partial x^2}+\frac{\partial^2 U}{\partial y^2},$$

or
$$\frac{\partial^2 U}{\partial x^2}+\frac{\partial^2 U}{\partial y^2}+\frac{\partial^2 U}{\partial z^2},$$

according as U is a function involving the one variable x, the two variables x, y, or the three variables x, y, z. The framework of rectangular Cartesian coordinates is always understood. In practice, U may also be a function of the further variable t, denoting time.

The equations with which we shall be dealing are

$$\nabla^2 U=0,$$

$$\nabla^2 U=\frac{1}{a}\frac{\partial U}{\partial t} \qquad (a \text{ constant}),$$

$$\nabla^2 U=\frac{1}{a^2}\frac{\partial^2 U}{\partial t^2} \qquad (a \text{ constant}).$$

The equation $\nabla^2 U=0$ is called LAPLACE'S EQUATION, and the symbol

$$\nabla^2 \equiv \frac{\partial^2}{\partial x^2}+\frac{\partial^2}{\partial y^2}+\frac{\partial^2}{\partial z^2}$$

is sometimes known as LAPLACE'S OPERATOR

Our first task is to transform these equations so as to express them in terms of other systems of coordinates, such as cylindrical or spherical polar coordinates.

2. The geometry of a transformation. Suppose that x, y, z are transformed to new variables u, v, w by means of relations

$$x=x(u,v,w),\quad y=y(u,v,w),\quad z=z(u,v,w),$$

where $x(u,v,w)$, $y(u,v,w)$, $z(u,v,w)$ are single-valued functions of u, v, w.

When u, v, w have fixed values u_0, v_0, w_0, there is determined a point $P_0(x_0, y_0, z_0)$; when one of them, say u, is variable while the others have fixed values v_0, w_0, the point so determined lies on a certain curve; when two of them, say v, w, are variable while the other has a fixed value u_0, the point so determined lies on a certain surface. We speak of THE SURFACE $u=u_0$, and of THE CURVE $v=v_0$, $w=w_0$ which is the intersection of the two surfaces $v=v_0$, $w=w_0$.

For example, we may identify u, v, w with the variables r, θ, ϕ of spherical polar coordinates, the three relations then being

$$x=r\sin\theta\cos\phi,\quad y=r\sin\theta\sin\phi,\quad z=r\cos\theta.$$

The surface $r=a$ is a sphere, the surface $\theta=\alpha$ is a cone (a plane when $\theta=\frac{1}{2}\pi$), the surface $\phi=\beta$ is a plane. The curve $r=a$, $\theta=\alpha$ is a circle, the curve $r=a$, $\phi=\beta$ is a circle, the curve $\theta=\alpha$, $\phi=\beta$ is a straight line.

It is assumed that the functions $x(u,v,w)$, $y(u,v,w)$, $z(u,v,w)$ are of such a nature that they can be solved to give u, v, w as single-valued functions of x, y, z, possibly for restricted ranges of values of the variables. For example, the solution in spherical polars is

$$r=\sqrt{(x^2+y^2+z^2)},\quad \theta=\tan^{-1}\frac{\sqrt{(x^2+y^2)}}{z},\quad \phi=\tan^{-1}\frac{y}{x},$$

and is unique for the region

$$r>0,\quad 0\leqslant\theta\leqslant\pi,\quad -\pi<\phi\leqslant\pi.$$

(Note that $\sin\theta=\dfrac{\sqrt{(x^2+y^2)}}{r}$, so that it is necessarily positive; hence the greater restriction on θ as compared with ϕ.)

We write the solution for u, v, w in the form

$$u=u(x,y,z),\quad v=v(x,y,z),\quad w=w(x,y,z).$$

The position of a point P may therefore be indicated by either of the alternative notations (x,y,z) or (u,v,w).

It will usually be assumed without further mention that the partial differential coefficients arising in the work are all continuous, and themselves possessing such partial differential coefficients as may be required.

Our immediate problem is *to find an expression for the tangent plane to the surface*

$$u(x, y, z) = u_0$$

at the point $P_0(x_0, y_0, z_0)$ *on it.*

Suppose that $P(x_0+h, y_0+j, z_0+k)$ is a point on the surface, lying close to P_0, so that

$$\begin{aligned} u(x_0+h, y_0+j, z_0+k) &= u_0 \\ &= u(x_0, y_0, z_0). \end{aligned}$$

By Taylor's theorem (Volume III, p. 57), there exists a number θ in the interval $(0, 1)$ such that

$$\begin{aligned} &h\frac{\partial}{\partial x}u(x_0+\theta h, y_0+\theta j, z_0+\theta k) \\ &\qquad +j\frac{\partial}{\partial y}u(x_0+\theta h, y_0+\theta j, z_0+\theta k) \\ &\qquad\qquad +k\frac{\partial}{\partial z}u(x_0+\theta h, y_0+\theta j, z_0+\theta k) = 0. \end{aligned}$$

Now let $Q(x, y, z)$ be an arbitrary point of the line P_0P. Then the increments h, j, k and $x-x_0$, $y-y_0$, $z-z_0$ are proportional, so that

$$\frac{h}{x-x_0} = \frac{j}{y-y_0} = \frac{k}{z-z_0}.$$

Hence the 'Taylor' relation becomes

$$\sum_{x,y,z} (x-x_0)\frac{\partial}{\partial x}u(x_0+\theta h, y_0+\theta j, z_0+\theta k) = 0.$$

This relation is, so far, *exact*, and we now interpret it when the point P of the surface moves to approach P_0 as a limiting position. If the partial differential coefficients are continuous, then, as $h, j, k \to 0$, the equation assumes the form

$$(x-x_0)\frac{\partial u}{\partial x_0} + (y-y_0)\frac{\partial u}{\partial y_0} + (z-z_0)\frac{\partial u}{\partial z_0} = 0,$$

where, for example, $\dfrac{\partial u}{\partial x_0}$ means $\dfrac{\partial}{\partial x}u(x, y, z)$ with x, y, z replaced by x_0, y_0, z_0 after differentiation.

This is the equation of a plane, having the property that it contains each line through P_0 which is a tangent there to the surface. It is called the TANGENT PLANE to the surface at P_0. Thus *the equation of the tangent plane at P_0 is*

$$(x-x_0)\frac{\partial u}{\partial x_0}+(y-y_0)\frac{\partial u}{\partial y_0}+(z-z_0)\frac{\partial u}{\partial z_0}=0.$$

COROLLARY. *The direction-cosines of the tangent plane are proportional to*

$$\left(\frac{\partial u}{\partial x_0}, \frac{\partial u}{\partial y_0}, \frac{\partial u}{\partial z_0}\right).$$

NOTE. A surface may have special points at which each of the differential coefficients $\partial u/\partial x_0$, $\partial u/\partial y_0$, $\partial u/\partial z_0$ vanishes. Such is, for instance, a *cone*, where they are zero at the vertex. We regard this possibility as excluded in the subsequent work.

3. Some general formulae. By direct application of the 'chain rule' (Volume III, pp. 25–7) we may establish three sets of equations of which the set

$$\frac{\partial x}{\partial u}\frac{\partial u}{\partial x}+\frac{\partial x}{\partial v}\frac{\partial v}{\partial x}+\frac{\partial x}{\partial w}\frac{\partial w}{\partial x}=1,$$

$$\frac{\partial x}{\partial u}\frac{\partial u}{\partial y}+\frac{\partial x}{\partial v}\frac{\partial v}{\partial y}+\frac{\partial x}{\partial w}\frac{\partial w}{\partial y}=0,$$

$$\frac{\partial x}{\partial u}\frac{\partial u}{\partial z}+\frac{\partial x}{\partial v}\frac{\partial v}{\partial z}+\frac{\partial x}{\partial w}\frac{\partial w}{\partial z}=0$$

is typical. We denote by J the Jacobian

$$J\equiv\frac{\partial(x, y, z)}{\partial(u, v, w)}\equiv\begin{vmatrix} \frac{\partial x}{\partial u} & \frac{\partial x}{\partial v} & \frac{\partial x}{\partial w} \\ \frac{\partial y}{\partial u} & \frac{\partial y}{\partial v} & \frac{\partial y}{\partial w} \\ \frac{\partial z}{\partial u} & \frac{\partial z}{\partial v} & \frac{\partial z}{\partial w} \end{vmatrix},$$

and by $X_{(u)}, X_{(v)}, \ldots, Z_{(w)}$ the cofactors of $x_u, x_v, \ldots, z_w$ in the determinant, so that, for example,

$$X_{(u)}\equiv\frac{\partial y}{\partial v}\frac{\partial z}{\partial w}-\frac{\partial y}{\partial w}\frac{\partial z}{\partial v},$$

$$X_{(v)}\equiv\frac{\partial y}{\partial w}\frac{\partial z}{\partial u}-\frac{\partial y}{\partial u}\frac{\partial z}{\partial w},$$

and so on. Then (Volume III, p. 98) it is known that

$$J^{-1} = \frac{\partial(u, v, w)}{\partial(x, y, z)} \equiv \begin{vmatrix} \frac{\partial u}{\partial x} & \frac{\partial u}{\partial y} & \frac{\partial u}{\partial z} \\ \frac{\partial v}{\partial x} & \frac{\partial v}{\partial y} & \frac{\partial v}{\partial z} \\ \frac{\partial w}{\partial x} & \frac{\partial w}{\partial y} & \frac{\partial w}{\partial z} \end{vmatrix},$$

and we denote the cofactors of $u_x, u_y, \ldots, w_z$ by the letters $U_{(x)}$, $U_{(y)}, \ldots, W_{(z)}$.

The set of equations may be solved for $\partial x/\partial u$, $\partial x/\partial v$, $\partial x/\partial w$ by the usual methods. Thus we find $\partial x/\partial u$ by multiplying by $U_{(x)}$, $U_{(y)}$, $U_{(z)}$ and adding, so that

$$J^{-1}\frac{\partial x}{\partial u} = U_{(x)}.$$

Similarly

$$J^{-1}\frac{\partial x}{\partial v} = V_{(x)},$$

$$\cdots\cdots\cdots\cdots\cdots\cdots$$

$$J^{-1}\frac{\partial z}{\partial w} = W_{(z)}.$$

WARNING. Although $X_{(u)}$ might be written in Jacobian notation in the form

$$X_{(u)} = \frac{\partial(y, z)}{\partial(v, w)}$$

and $U_{(x)}$ in the form

$$U_{(x)} = \frac{\partial(v, w)}{\partial(y, z)},$$

it is **not true** that

$$X_{(u)} = 1/U_{(x)}.$$

The presence of the further variables x, u upsets the proof used for the case when y, z, v, w are the only variables involved. It is doubtful whether such uses of the Jacobian notation are proper, and we have avoided them.

4. Mutually orthogonal systems. Two surfaces

$$u(x, y, z) = u_0, \quad v(x, y, z) = v_0$$

are said to cut ORTHOGONALLY at a point P_0 if their tangent planes there are perpendicular. By the Corollary at the end of § 2, the condition for this is

$$\frac{\partial u}{\partial x_0}\frac{\partial v}{\partial x_0} + \frac{\partial u}{\partial y_0}\frac{\partial v}{\partial y_0} + \frac{\partial u}{\partial z_0}\frac{\partial v}{\partial z_0} = 0.$$

Three surfaces

$$u(x, y, z) = u_0, \quad v(x, y, z) = v_0, \quad w(x, y, z) = w_0$$

are said to be MUTUALLY ORTHOGONAL at a point P_0 if their tangent planes taken in pairs are perpendicular. The conditions for orthogonality at a general point are (dropping suffixes)

$$\frac{\partial v}{\partial x}\frac{\partial w}{\partial x} + \frac{\partial v}{\partial y}\frac{\partial w}{\partial y} + \frac{\partial v}{\partial z}\frac{\partial w}{\partial z} = 0,$$

$$\frac{\partial w}{\partial x}\frac{\partial u}{\partial x} + \frac{\partial w}{\partial y}\frac{\partial u}{\partial y} + \frac{\partial w}{\partial z}\frac{\partial u}{\partial z} = 0,$$

$$\frac{\partial u}{\partial x}\frac{\partial v}{\partial x} + \frac{\partial u}{\partial y}\frac{\partial v}{\partial y} + \frac{\partial u}{\partial z}\frac{\partial v}{\partial z} = 0.$$

One or two deductions may be made at once from these relations. Solving the second and third equations for $\frac{\partial u}{\partial x} : \frac{\partial u}{\partial y} : \frac{\partial u}{\partial z}$, we have

$$\frac{\dfrac{\partial u}{\partial x}}{\dfrac{\partial v}{\partial y}\dfrac{\partial w}{\partial z} - \dfrac{\partial v}{\partial z}\dfrac{\partial w}{\partial y}} = \frac{\dfrac{\partial u}{\partial y}}{\dfrac{\partial v}{\partial z}\dfrac{\partial w}{\partial x} - \dfrac{\partial v}{\partial x}\dfrac{\partial w}{\partial z}} = \frac{\dfrac{\partial u}{\partial z}}{\dfrac{\partial v}{\partial x}\dfrac{\partial w}{\partial y} - \dfrac{\partial v}{\partial y}\dfrac{\partial w}{\partial x}},$$

or, in the notation of § 3,

$$\frac{\dfrac{\partial u}{\partial x}}{U_{(x)}} = \frac{\dfrac{\partial u}{\partial y}}{U_{(y)}} = \frac{\dfrac{\partial u}{\partial z}}{U_{(z)}}.$$

The formulae at the end of § 3 now give the relations

$$\frac{\dfrac{\partial u}{\partial x}}{\dfrac{\partial x}{\partial u}} = \frac{\dfrac{\partial u}{\partial y}}{\dfrac{\partial y}{\partial u}} = \frac{\dfrac{\partial u}{\partial z}}{\dfrac{\partial z}{\partial u}}.$$

If we write (for reference in the next paragraph)

$$\left(\frac{\partial x}{\partial u}\right)^2 + \left(\frac{\partial y}{\partial u}\right)^2 + \left(\frac{\partial z}{\partial u}\right)^2 = h_1^2,$$

$$\left(\frac{\partial x}{\partial v}\right)^2 + \left(\frac{\partial y}{\partial v}\right)^2 + \left(\frac{\partial z}{\partial v}\right)^2 = h_2^2,$$

$$\left(\frac{\partial x}{\partial w}\right)^2 + \left(\frac{\partial y}{\partial w}\right)^2 + \left(\frac{\partial z}{\partial w}\right)^2 = h_3^2,$$

where h_1, h_2, h_3 may be taken to be positive, if required, and also denote the three ratios given above by λ, so that

$$\frac{\partial u}{\partial x}=\lambda\frac{\partial x}{\partial u},\quad \frac{\partial u}{\partial y}=\lambda\frac{\partial y}{\partial u},\quad \frac{\partial u}{\partial z}=\lambda\frac{\partial z}{\partial u},$$

then the relation
$$\frac{\partial u}{\partial x}\frac{\partial x}{\partial u}+\frac{\partial u}{\partial y}\frac{\partial y}{\partial u}+\frac{\partial u}{\partial z}\frac{\partial z}{\partial u}=1$$

shows that
$$\lambda h_1^2=1.$$

Hence we have the formulae

$$\frac{\partial u}{\partial x}=\frac{1}{h_1^2}\frac{\partial x}{\partial u},\quad \frac{\partial u}{\partial y}=\frac{1}{h_1^2}\frac{\partial y}{\partial u},\quad \frac{\partial u}{\partial z}=\frac{1}{h_1^2}\frac{\partial z}{\partial u},$$

with similar results for v (with h_2^2) and w (with h_3^2).

COROLLARY. It follows at once that (in this case of orthogonality)

$$\left(\frac{\partial u}{\partial x}\right)^2+\left(\frac{\partial u}{\partial y}\right)^2+\left(\frac{\partial u}{\partial z}\right)^2=\frac{1}{h_1^2},$$

$$\left(\frac{\partial v}{\partial x}\right)^2+\left(\frac{\partial v}{\partial y}\right)^2+\left(\frac{\partial v}{\partial z}\right)^2=\frac{1}{h_2^2},$$

$$\left(\frac{\partial w}{\partial x}\right)^2+\left(\frac{\partial w}{\partial y}\right)^2+\left(\frac{\partial w}{\partial z}\right)^2=\frac{1}{h_3^2}.$$

The importance which attaches to the functions h_1, h_2, h_3 is seen by the theorem that, FOR ORTHOGONAL RELATIONS, *the differentials dx, dy, dz are connected with du, dv, dw by means of the formula*

$$dx^2+dy^2+dz^2=h_1^2du^2+h_2^2dv^2+h_3^2dw^2.$$

The proof follows by squaring and adding the three relations such as

$$dx=\frac{\partial x}{\partial u}du+\frac{\partial x}{\partial v}dv+\frac{\partial x}{\partial w}dw,$$

and noting that, for example,

$$\begin{aligned}&\frac{\partial x}{\partial v}\frac{\partial x}{\partial w}+\frac{\partial y}{\partial v}\frac{\partial y}{\partial w}+\frac{\partial z}{\partial v}\frac{\partial z}{\partial w}\\ &=h_2^2h_3^2\left(\frac{\partial v}{\partial x}\frac{\partial w}{\partial x}+\frac{\partial v}{\partial y}\frac{\partial w}{\partial y}+\frac{\partial v}{\partial z}\frac{\partial w}{\partial z}\right)\\ &=0.\end{aligned}$$

A useful formula is obtained by multiplying the two equivalent determinants

$$\frac{1}{J} \equiv \begin{vmatrix} u_x & u_y & u_z \\ v_x & v_y & v_z \\ w_x & w_y & w_z \end{vmatrix}, \quad \frac{1}{J} \equiv \begin{vmatrix} u_x & v_x & w_x \\ u_y & v_y & w_y \\ u_z & v_z & w_z \end{vmatrix}.$$

By the standard ('matrix') rule of multiplication,

$$\frac{1}{J^2} = \begin{vmatrix} u_x^2 + u_y^2 + u_z^2 & u_x v_x + u_y v_y + u_z v_z & u_x w_x + u_y w_y + u_z w_z \\ v_x u_x + v_y u_y + v_z u_z & v_x^2 + v_y^2 + v_z^2 & v_x w_x + v_y w_y + v_z w_z \\ w_x u_x + w_y u_y + w_z u_z & w_x v_x + w_y v_y + w_z v_z & w_x^2 + w_y^2 + w_z^2 \end{vmatrix}$$

$$= \begin{vmatrix} h_1^{-2} & 0 & 0 \\ 0 & h_2^{-2} & 0 \\ 0 & 0 & h_3^{-2} \end{vmatrix},$$

in virtue of the orthogonality relations. Hence

$$J = \pm h_1 h_2 h_3.$$

The sign of J may be taken as positive should we wish, since the interchange of the names of two of the functions $x(u, v, w)$, $y(u, v, w)$, $z(u, v, w)$ would interchange the values of two rows of J and so change the sign if required. With the convention J positive (and h_1, h_2, h_3 positive) the formula is

$$J = +h_1 h_2 h_3.$$

(The assumed constancy in sign requires that J is never zero here. Compare Volume III, pp. 90–5.)

5. The transformation of $\nabla^2 U$. Suppose that x, y, z are transformed to new variables u, v, w by means of the transformation

$$x = x(u, v, w), \quad y = y(u, v, w), \quad z = z(u, v, w).$$

We restrict the discussion to the case when the three surfaces $u = \text{const.}$, $v = \text{const.}$, $w = \text{const.}$ are MUTUALLY ORTHOGONAL.

By direct differentiation, we have the relation*

$$U_x = U_u u_x + U_v v_x + U_w w_x,$$

so that, on differentiating again,

$$U_{xx} = U_u u_{xx} + U_v v_{xx} + U_w w_{xx} + \sum_{u, v, w} (U_{uu} u_x + U_{uv} v_x + U_{uw} w_x) u_x.$$

* $U_x \equiv \dfrac{\partial U}{\partial x}$, and so on.

It follows, in virtue of the orthogonality conditions of § 4, that

$$\nabla^2 U = U_u \nabla^2 u + U_v \nabla^2 v + U_w \nabla^2 w + \sum_{u,\,v,\,w} U_{uu}\{u_x^2 + u_y^2 + u_z^2\}.$$

Consider the part of this expression which involves differentiations with respect to u, namely

$$A \equiv U_u \nabla^2 u + U_{uu}(u_x^2 + u_y^2 + u_z^2).$$

We wish to convert the differentiations of u with respect to x, y, z into differentiations of x, y, z with respect to u, v, w, since this is the more natural process when x, y, z are given as functions of u, v, w.

We have already proved (p. 225) that

$$u_x^2 + u_y^2 + u_z^2 = \frac{1}{h_1^2},$$

so we pass to the evaluation of $\nabla^2 u$. Since (p. 225)

$$\frac{\partial u}{\partial x} = \frac{1}{h_1^2}\frac{\partial x}{\partial u},$$

it follows that
$$\frac{\partial^2 u}{\partial x^2} = -\frac{2}{h_1^3}\frac{\partial h_1}{\partial x}\frac{\partial x}{\partial u} + \frac{1}{h_1^2}\frac{\partial}{\partial x}\left(\frac{\partial x}{\partial u}\right).$$

But
$$\frac{\partial x}{\partial u} = J\left(\frac{\partial v}{\partial y}\frac{\partial w}{\partial z} - \frac{\partial v}{\partial z}\frac{\partial w}{\partial y}\right),$$

so that

$$\frac{\partial}{\partial x}\left(\frac{\partial x}{\partial u}\right) = \frac{\partial J}{\partial x}\left(\frac{\partial v}{\partial y}\frac{\partial w}{\partial z} - \frac{\partial v}{\partial z}\frac{\partial w}{\partial y}\right) + J\frac{\partial}{\partial x}\left(\frac{\partial v}{\partial y}\frac{\partial w}{\partial z} - \frac{\partial v}{\partial z}\frac{\partial w}{\partial y}\right).$$

Substituting and adding for x, y, z, we obtain the formula

$$\begin{aligned}
\nabla^2 u &= -\frac{2}{h_1^3}\left(\frac{\partial h_1}{\partial x}\frac{\partial x}{\partial u} + \frac{\partial h_1}{\partial y}\frac{\partial y}{\partial u} + \frac{\partial h_1}{\partial z}\frac{\partial z}{\partial u}\right) \\
&\quad + \frac{1}{h_1^2}\left\{\frac{\partial(J, v, w)}{\partial(x, y, z)} + J\,.\,0\right\} \\
&= -\frac{2}{h_1^3}\frac{\partial h_1}{\partial u} + \frac{1}{h_1^2}\frac{\partial(J, v, w)}{\partial(x, y, z)} \\
&= -\frac{2}{h_1^3}\frac{\partial h_1}{\partial u} + \frac{1}{h_1^2}\frac{\partial(J, v, w)}{\partial(u, v, w)}\frac{\partial(u, v, w)}{\partial(x, y, z)} \\
&= -\frac{2}{h_1^3}\frac{\partial h_1}{\partial u} + \frac{1}{h_1^2}\frac{\partial J}{\partial u}\cdot\frac{1}{J} \\
&= -\frac{2}{h_1^3}\frac{\partial h_1}{\partial u} + \frac{1}{h_1^3 h_2 h_3}\frac{\partial(h_1 h_2 h_3)}{\partial u},
\end{aligned}$$

whatever convention of sign is adopted for J (p. 226).

Thus

$$A \equiv -\frac{2}{h_1^3}\frac{\partial h_1}{\partial u}\frac{\partial U}{\partial u} + \frac{1}{h_1^3 h_2 h_3}\frac{\partial (h_1 h_2 h_3)}{\partial u}\frac{\partial U}{\partial u} + \frac{1}{h_1^2}\frac{\partial^2 U}{\partial u^2}$$

$$= \frac{1}{h_1^2}\frac{\partial^2 U}{\partial u^2} + \frac{1}{h_1^3}\left\{-\frac{\partial h_1}{\partial u} + \frac{h_1}{h_2 h_3}\frac{\partial (h_2 h_3)}{\partial u}\right\}\frac{\partial U}{\partial u}$$

$$= \frac{1}{h_1 h_2 h_3}\left\{\frac{h_2 h_3}{h_1}\frac{\partial^2 U}{\partial u^2} + \frac{\partial}{\partial u}\left(\frac{h_2 h_3}{h_1}\right)\frac{\partial U}{\partial u}\right\}$$

$$= \frac{1}{h_1 h_2 h_3}\frac{\partial}{\partial u}\left\{\frac{h_2 h_3}{h_1}\frac{\partial U}{\partial u}\right\}.$$

We therefore have the following rule:

If x, y, z are transformed to u, v, w by means of the ORTHOGONAL *relations*

$$x = x(u, v, w), \quad y = y(u, v, w), \quad z = z(u, v, w),$$

so that

$$dx^2 + dy^2 + dz^2 = h_1^2 du^2 + h_2^2 dv^2 + h_3^2 dw^2,$$

then the transformation of $\nabla^2 U$ is given by the formula

$$\nabla^2 U \equiv \frac{1}{h_1 h_2 h_3}\left\{\frac{\partial}{\partial u}\left(\frac{h_2 h_3}{h_1}\frac{\partial U}{\partial u}\right) + \frac{\partial}{\partial v}\left(\frac{h_3 h_1}{h_2}\frac{\partial U}{\partial v}\right) + \frac{\partial}{\partial w}\left(\frac{h_1 h_2}{h_3}\frac{\partial U}{\partial w}\right)\right\}.$$

NOTE. Mr F. Bowman has derived, in the *Mathematical Gazette*, XXX (1941), p. 51, the more general formula

$$\nabla^2 U \equiv \frac{1}{J}\sum_{u,v,w}\frac{\partial}{\partial u}\left\{\frac{1}{J}\left|\begin{matrix} U_u & U_v & U_w \\ h & b & f \\ g & f & c \end{matrix}\right|\right\},$$

where a, b, c are written for our h_1^2, h_2^2, h_3^2, and where

$$f \equiv x_v x_w + y_v y_w + z_v z_w,$$

with similar definitions for g and h. Those familiar with matrix technique will find a simpler exposition in *Elementary Matrices*, by R. A. Frazer, W. J. Duncan and A. R. Collar, p. 51 (Cambridge University Press, 1946).

The formula which we have given is often derived neatly from a result known as *Gauss's theorem*; but, of course, that has to be proved first. It is also possible that the neatness of some derivations arises from the neglect of detailed consideration of the volumes or areas of curvilinear 'boxes'. At any rate, the treatment here aims at rigour founded on elementary principles. But knowledge of the

'Gauss' method is essential for those who desire physical insight into what is involved in applications, and a text-book (say on Vector Methods) should be consulted.

COROLLARY. *The plane transformation of* $\nabla^2 U$. Suppose, as a particular case of the work of this paragraph, that U is a function of x, y only, and that the transformation is

$$x = x(u, v), \quad y = y(u, v), \quad z = w,$$

so that x, y are functions of u, v only. The surfaces $u = \text{const.}$, $v = \text{const.}$ are cylinders of which the sections by the planes $z = \text{const.}$, or $w = \text{const.}$, are the orthogonal sections, and the condition for these cylinders to cut orthogonally is

$$\frac{\partial u}{\partial x}\frac{\partial v}{\partial x} + \frac{\partial u}{\partial y}\frac{\partial v}{\partial y} = 0.$$

Since u, v are expressible as functions of x, y only, it follows that $\dfrac{\partial u}{\partial z} = \dfrac{\partial v}{\partial z} = 0$; further $\dfrac{\partial w}{\partial x} = 0$, $\dfrac{\partial w}{\partial y} = 0$. The three orthogonality conditions of §4 (p. 224) are therefore all satisfied, and so the subsequent analysis remains valid.

The functions h_1^2, h_2^2, h_3^2 are given, from their definition, by the relations

$$h_1^2 = \left(\frac{\partial x}{\partial u}\right)^2 + \left(\frac{\partial y}{\partial u}\right)^2,$$

$$h_2^2 = \left(\frac{\partial x}{\partial v}\right)^2 + \left(\frac{\partial y}{\partial v}\right)^2,$$

$$h_3^2 = 1.$$

Hence *the formula of transformation is*

$$\frac{\partial^2 U}{\partial x^2} + \frac{\partial^2 U}{\partial y^2} = \frac{1}{h_1 h_2}\left\{\frac{\partial}{\partial u}\left(\frac{h_2}{h_1}\frac{\partial U}{\partial u}\right) + \frac{\partial}{\partial v}\left(\frac{h_1}{h_2}\frac{\partial U}{\partial v}\right)\right\}.$$

EXAMPLES I

1. Prove that, in terms of polar coordinates $x = r\cos\theta$, $y = r\sin\theta$,

$$\frac{\partial^2 U}{\partial x^2} + \frac{\partial^2 U}{\partial y^2} = \frac{\partial^2 U}{\partial r^2} + \frac{1}{r}\frac{\partial U}{\partial r} + \frac{1}{r^2}\frac{\partial^2 U}{\partial \theta^2}.$$

2. Prove that, in terms of cylindrical coordinates $x = \rho\cos\phi$, $y = \rho\sin\phi$,

$$\frac{\partial^2 U}{\partial x^2} + \frac{\partial^2 U}{\partial y^2} + \frac{\partial^2 U}{\partial z^2} = \frac{\partial^2 U}{\partial \rho^2} + \frac{1}{\rho}\frac{\partial U}{\partial \rho} + \frac{1}{\rho^2}\frac{\partial^2 U}{\partial \phi^2} + \frac{\partial^2 U}{\partial z^2}.$$

3. Prove that, in terms of spherical polar coordinates

$$x = r\sin\theta\cos\phi, \quad y = r\sin\theta\sin\phi, \quad z = r\cos\theta,$$

$$\nabla^2 U \equiv \frac{1}{r^2}\frac{\partial}{\partial r}\left(r^2\frac{\partial U}{\partial r}\right) + \frac{1}{r^2\sin\theta}\frac{\partial}{\partial\theta}\left(\sin\theta\frac{\partial U}{\partial\theta}\right) + \frac{1}{r^2\sin^2\theta}\frac{\partial^2 U}{\partial\phi^2}$$

$$\equiv \frac{\partial^2 U}{\partial r^2} + \frac{2}{r}\frac{\partial U}{\partial r} + \frac{1}{r^2}\frac{\partial^2 U}{\partial\theta^2} + \frac{1}{r^2}\cot\theta\frac{\partial U}{\partial\theta} + \frac{1}{r^2\sin^2\theta}\frac{\partial^2 U}{\partial\phi^2}.$$

4. Determine the general form of a function V which depends only on distance from a given line in space and which satisfies the equation $\nabla^2 V = 0$.

CHAPTER XXX

'LAPLACE' EQUATIONS

(i) SOLUTION BY SEPARATION OF VARIABLES

The equations listed at the start of the preceding chapter (p. 219) can be solved in a bewildering variety of ways. One basic method, which can be applied widely, is to find, in the first instance, solutions in the form of products of functions each involving only one of the independent variables.

The work of this section shows how the method can be applied in a number of typical instances.

1. The equation of heat conduction. The flow of heat in a uniform body is governed by the equation

$$\nabla^2 V = \frac{1}{h^2}\frac{\partial V}{\partial t},$$

where V is the temperature at time t at the point (x, y, z) and h is a constant depending on the nature of the material. In particular, the equation for a thin rod is

$$\frac{\partial^2 V}{\partial x^2} = \frac{1}{h^2}\frac{\partial V}{\partial t}.$$

We are to solve this equation by a method which *separates* V into a part involving x only and a part involving t only; that is, we consider whether there are solutions of the form

$$V = XT,$$

where X is a function of x only and T a function of t only. If so, then the equation is

$$T\frac{d^2X}{dx^2} = \frac{1}{h^2}X\frac{dT}{dt},$$

or

$$\frac{1}{X}\frac{d^2X}{dx^2} = \frac{1}{h^2T}\frac{dT}{dt}.$$

Now the left-hand side is a function of x only, and the right-hand side is a function of t only; and they can be equal only by being constant, a function of neither. Denote this constant by the letter k. Thus

$$\frac{d^2X}{dx^2} = kX,$$

$$\frac{dT}{dt} = h^2kT.$$

The nature of the solution depends on the sign of k. For instance, solutions periodic in x may be obtained by giving k the negative value $-a^2$. Then

$$X = A\cos ax + B\sin ax,$$

$$T = Ce^{-a^2h^2t},$$

so that (absorbing C into the arbitrary constants A, B) a solution of the given equation is exhibited in the form

$$V = (A\cos ax + B\sin ax)\,e^{-a^2h^2t}.$$

Suppose, for example, that the ends $x=0$, $x=l$ of the rod are kept at zero temperature. Then there is a solution

$$V = B\sin ax\, e^{-a^2h^2t},$$

where $$al = n\pi \quad (n \text{ an integer}).$$

Thus the constant a assumes the form $n\pi/l$, so that

$$V = B\sin\frac{n\pi x}{l}e^{-n^2\pi^2h^2t/l^2}.$$

If it is known that, say, at time $t=0$ the temperature is distributed along the rod in accordance with a given Fourier series

$$V_0 = \sum_{n=1}^{\infty} \alpha_n \sin\frac{n\pi x}{l},$$

then the temperature at time t is given by the relation

$$V = \sum_{n=1}^{\infty} \alpha_n \sin\frac{n\pi x}{l}\, e^{-n^2\pi^2h^2t/l^2},$$

for this function satisfies the differential equation and also the given initial condition.

2. The solution of Laplace's equation in plane polar coordinates. Laplace's equation is

$$\frac{\partial^2 U}{\partial r^2} + \frac{1}{r}\frac{\partial U}{\partial r} + \frac{1}{r^2}\frac{\partial^2 U}{\partial \theta^2} = 0,$$

and we require, for the method of separation of variables, solutions in the form

$$U = R\Theta,$$

where R is a function of r only and Θ a function of θ only. Thus

$$\Theta\left\{\frac{d^2R}{dr^2}+\frac{1}{r}\frac{dR}{dr}\right\}+\frac{R}{r^2}\frac{d^2\Theta}{d\theta^2}=0,$$

or

$$\frac{r^2}{R}\left\{\frac{d^2R}{dr^2}+\frac{1}{r}\frac{dR}{dr}\right\}=-\frac{1}{\Theta}\frac{d^2\Theta}{d\theta^2}.$$

Since the left-hand side involves only r, and the right-hand side only θ, each is constant. We may, for illustration, consider solutions which are periodic in θ, so that the constant is positive, say n^2. Then

$$\Theta = A\cos n\theta + B\sin n\theta.$$

The equation for R is

$$r^2\frac{d^2R}{dr^2}+r\frac{dR}{dr}-n^2R=0,$$

and the standard substitution (p. 52) $r=e^v$ gives the equation

$$\frac{d^2R}{dv^2}-n^2R=0$$

with solution

$$\begin{aligned}R &= Pe^{nv}+Qe^{-nv}\\ &= Pr^n+Qr^{-n},\end{aligned}$$

where P, Q are arbitrary constants.

A solution of the given equation may thus be taken in the form

$$U = ar^n\cos n\theta + br^n\sin n\theta + cr^{-n}\cos n\theta + dr^{-n}\sin n\theta.$$

A more general solution is obtained by adding any number of such solutions for different values of n; this is easily verified by substituting in the given equation, when each contribution vanishes separately. In particular, by giving n all positive integral powers, we obtain solutions in the form of a series

$$U=\sum_{n=1}^{\infty}(a_nr^n\cos n\theta+b_nr^n\sin n\theta+c_nr^{-n}\cos n\theta+d_nr^{-n}\sin n\theta).$$

3. The wave equation in two dimensions. Consider next the equation

$$\frac{\partial^2U}{\partial x^2}+\frac{\partial^2U}{\partial y^2}=\frac{1}{c^2}\frac{\partial^2U}{\partial t^2},$$

for which we require solutions in the form

$$U = XYT,$$

where X, Y, T are functions of x, y, t respectively. By substitution in the given equation, we have

$$YT\frac{d^2X}{dx^2} + XT\frac{d^2Y}{dy^2} = \frac{1}{c^2}XY\frac{d^2T}{dt^2},$$

or

$$\frac{1}{X}\frac{d^2X}{dx^2} + \frac{1}{Y}\frac{d^2Y}{dy^2} = \frac{1}{c^2T}\frac{d^2T}{dt^2}.$$

Each side must be constant, say (for solutions periodic in t) $-n^2$. Then

$$T = A\cos nct + B\sin nct;$$

also X, Y satisfy the relation

$$\frac{1}{X}\frac{d^2X}{dx^2} = -\frac{1}{Y}\frac{d^2Y}{dy^2} - n^2,$$

where, again, each side must be constant. If, for example, we search for solutions periodic in X, we may take that constant as $-m^2$; then

$$X = P\cos mx + Q\sin mx$$

and

$$\frac{d^2Y}{dy^2} + (n^2 - m^2)\,Y = 0.$$

The solutions in Y are periodic if $n^2 > m^2$ and exponential if $n^2 < m^2$. Suppose that we seek solutions which tend to zero as y tends to infinity. Then we take $n^2 < m^2$ and obtain the solution

$$Y = Ce^{-\sqrt{(m^2-n^2)}y}.$$

Thus for a solution which, say, is periodic in t and is stationary $(\partial U/\partial t = 0)$ at time $t = 0$, which is periodic in x and vanishes at $x = 0$, and which tends to zero as y tends to infinity, we may adopt the form

$$U = A\sin mx\, e^{-\sqrt{(m^2-n^2)}y}\cos nct,$$

where m, n are any constants such that $m^2 > n^2$. If, for example, conditions are such that the vibration at any point has a given period $2\pi/c$, then we take $n = 1$ and obtain the solution

$$U = A\sin mx\, e^{-\sqrt{(m^2-1)}y}\cos ct;$$

and if, further, U is always zero along the line $x = l$, then m must be of the form $k\pi/l$. A solution under all these conditions can therefore be obtained in a general form

$$U = \sum_k A \sin\frac{k\pi x}{l} \exp\left[-\sqrt{\left(\frac{k^2\pi^2}{l^2}-1\right)}y\right]\cos ct,$$

summed for values of k greater than l/π (for which the expression under the square root sign is positive).

The solutions hitherto obtained have involved only functions such as sines, cosines, exponentials and the like, with which we are already familiar. But other choices of coordinates involve functions essentially new, though we met them from a preliminary point of view while solving differential equations in the form of infinite series. The primary purpose of the next few paragraphs is to show how fresh functions become necessary, though more detailed study of their properties is reserved for later.

4. The equation $\frac{\partial^2 U}{\partial x^2}+\frac{\partial^2 U}{\partial y^2}+\frac{\partial^2 U}{\partial z^2}=0$ in spherical polar coordinates; Legendre's equation. The equation in spherical polar coordinates is (p. 230)

$$\frac{\partial^2 U}{\partial r^2}+\frac{2}{r}\frac{\partial U}{\partial r}+\frac{1}{r^2}\frac{\partial^2 U}{\partial \theta^2}+\frac{1}{r^2}\cot\theta\frac{\partial U}{\partial \theta}+\frac{1}{r^2\sin^2\theta}\frac{\partial^2 U}{\partial \phi^2}=0.$$

We seek solutions of the form

$$U = R\Theta\Phi,$$

where R is a function of r only, Θ of θ only, and Φ of ϕ only. On multiplying by r^2, substituting, and dividing by $R\Theta\Phi$, we obtain the equation

$$\frac{1}{R}\left(r^2\frac{d^2R}{dr^2}+2r\frac{dR}{dr}\right)+\frac{1}{\Theta}\left(\frac{d^2\Theta}{d\theta^2}+\cot\theta\frac{d\Theta}{d\theta}\right)+\frac{1}{\Phi\sin^2\theta}\frac{d^2\Phi}{d\phi^2}=0.$$

By an argument now familiar, we have a relation of the form

$$r^2\frac{d^2R}{dr^2}+2r\frac{dR}{dr}+kR=0,$$

where k is a constant. If we make the substitution

$$r = e^u,$$

then (compare p. 233) the equation becomes

$$\frac{d^2R}{du^2}+\frac{dR}{du}+kR=0.$$

This is a linear equation with constant coefficients, whose auxiliary equation (p. 32) is

$$p^2+p+k=0.$$

The roots of this equation have sum -1, so we may write them in the form n, $-(n+1)$, noting incidentally that

$$k=-n(n+1).$$

Then
$$R=Ae^{nu}+Be^{-(n+1)u}$$
$$=Ar^n+\frac{B}{r^{n+1}}.$$

We now have the relation

$$\frac{1}{\Theta}\left(\frac{d^2\Theta}{d\theta^2}+\cot\theta\frac{d\Theta}{d\theta}\right)+\frac{1}{\Phi\sin^2\theta}\frac{d^2\Phi}{d\phi^2}+n(n+1)=0,$$

or
$$\frac{1}{\Theta}\left(\frac{d^2\Theta}{d\theta^2}+\cot\theta\frac{d\Theta}{d\theta}\right)\sin^2\theta+n(n+1)\sin^2\theta=-\frac{1}{\Phi}\frac{d^2\Phi}{d\phi^2}.$$

Now the geometrical meaning for ϕ suggests that there may be useful solutions in which Φ is a periodic function of ϕ, and such solutions can be found by setting each side of this equation equal to m^2, so that

$$\frac{d^2\Phi}{d\phi^2}+m^2\Phi=0,$$

or
$$\Phi=C\cos m\phi+D\sin m\phi.$$

The equation for Θ is then

$$\frac{d^2\Theta}{d\theta^2}+\cot\theta\frac{d\Theta}{d\theta}+\left\{n(n+1)-\frac{m^2}{\sin^2\theta}\right\}\Theta=0.$$

The standard form in which this equation is usually given is obtained by means of the substitution

$$\mu=\cos\theta.$$

Then
$$\frac{d\Theta}{d\theta}=-\frac{d\Theta}{d\mu}\sin\theta$$

and
$$\frac{d^2\Theta}{d\theta^2}=\frac{d^2\Theta}{d\mu^2}\sin^2\theta-\frac{d\Theta}{d\mu}\cos\theta,$$

so that
$$\frac{d^2\Theta}{d\theta^2}+\cot\theta\frac{d\Theta}{d\theta}=\frac{d^2\Theta}{d\mu^2}\sin^2\theta-2\frac{d\Theta}{d\mu}\cos\theta$$
$$=(1-\mu^2)\frac{d^2\Theta}{d\mu^2}-2\mu\frac{d\Theta}{d\mu}$$
$$=\frac{d}{d\mu}\left\{(1-\mu^2)\frac{d\Theta}{d\mu}\right\}.$$

Hence the equation is
$$\frac{d}{d\mu}\left\{(1-\mu^2)\frac{d\Theta}{d\mu}\right\}+\left\{n(n+1)-\frac{m^2}{1-\mu^2}\right\}\Theta=0.$$

This equation is known as LEGENDRE'S ASSOCIATED EQUATION.

There are many problems in which solutions of the given equation
$$\frac{\partial^2 U}{\partial x^2}+\frac{\partial^2 U}{\partial y^2}+\frac{\partial^2 U}{\partial z^2}=0$$
in the form
$$U=R\Theta\Phi$$
are required to be independent of the variable ϕ, so that they are 'symmetrical about the z-axis'. The constant m must then be zero, and the equation for Θ is
$$\frac{d}{d\mu}\left\{(1-\mu^2)\frac{d\Theta}{d\mu}\right\}+n(n+1)\Theta=0.$$

This is called LEGENDRE'S EQUATION OF ORDER n.

If
$$\Theta\equiv\Theta_n(\mu)$$
is any solution of this equation, then solutions of the equation $\frac{\partial^2 U}{\partial x^2}+\frac{\partial^2 U}{\partial y^2}+\frac{\partial^2 U}{\partial z^2}=0$ symmetrical about the z-axis may be found as sums of terms of the type
$$r^n\Theta_n,\quad r^{-(n+1)}\Theta_n,$$
say
$$U=\sum_n\left(A_n r^n+\frac{B_n}{r^{n+1}}\right)\Theta_n,$$
where A_n, B_n are constants.

5. The equation $\frac{\partial^2 U}{\partial x^2}+\frac{\partial^2 U}{\partial y^2}+\frac{\partial^2 U}{\partial z^2}=0$ in cylindrical coordinates; Bessel's equation. The equation in cylindrical coordinates is (p. 229)
$$\frac{\partial^2 U}{\partial \rho^2}+\frac{1}{\rho}\frac{\partial U}{\partial \rho}+\frac{1}{\rho^2}\frac{\partial^2 U}{\partial \phi^2}+\frac{\partial^2 U}{\partial z^2}=0.$$

We seek solutions of the form

$$U = S\Phi Z,$$

where S is a function of ρ only, Φ of ϕ only, and Z of z only. Substitute and divide by $S\Phi Z$; then

$$\frac{1}{S}\frac{d^2S}{d\rho^2}+\frac{1}{S\rho}\frac{dS}{d\rho}+\frac{1}{\Phi\rho^2}\frac{d^2\Phi}{d\phi^2}=-\frac{1}{Z}\frac{d^2Z}{dz^2}.$$

As before (p. 234), each side is constant. For solutions in which, for example, Z assumes exponential form, that constant may be put equal to $-m^2$, so that

$$\frac{d^2Z}{dz^2}=m^2Z,$$

or

$$Z = Ae^{mz}+Be^{-mz}.$$

Then

$$\frac{1}{S}\frac{d^2S}{d\rho^2}+\frac{1}{S\rho}\frac{dS}{d\rho}+\frac{1}{\Phi\rho^2}\frac{d^2\Phi}{d\phi^2}=-m^2,$$

so that

$$\rho^2\left\{\frac{1}{S}\frac{d^2S}{d\rho^2}+\frac{1}{S\rho}\frac{dS}{d\rho}+m^2\right\}=-\frac{1}{\Phi}\frac{d^2\Phi}{d\phi^2}.$$

Once again each side is constant. Solutions which are, say, periodic in ϕ may be obtained by putting that constant equal to n^2, so that

$$\frac{d^2\Phi}{d\phi^2}+n^2\Phi=0,$$

or

$$\Phi = C\cos n\phi + D\sin n\phi.$$

The equation for S is then

$$\rho^2\frac{d^2S}{d\rho^2}+\rho\frac{dS}{d\rho}+m^2S\rho^2=n^2S.$$

The standard form of this equation is obtained by making the substitution

$$v = m\rho,$$

giving

$$v^2\frac{d^2S}{dv^2}+v\frac{dS}{dv}+(v^2-n^2)S=0.$$

This is a very famous equation, known as BESSEL'S EQUATION, and its solutions are called BESSEL FUNCTIONS OF ORDER n. These functions have many important properties to which many authors have devoted detailed study. Here we note that, if

$$S_n(v)$$

is a solution of Bessel's equation of order n, then solutions of the equation

$$\frac{\partial^2 U}{\partial x^2}+\frac{\partial^2 U}{\partial y^2}+\frac{\partial^2 U}{\partial z^2}=0$$

are obtained as sums of expressions of the form

$$e^{\pm mz}S_n(m\rho)\begin{Bmatrix}\cos n\phi\\ \sin n\phi\end{Bmatrix}.$$

(ii) OTHER METHODS OF SOLUTION

We do not consider in these volumes the detailed methods available for the solution of partial differential equations, but one or two of them may be illustrated incidentally by reference to the particular group of equations with which we are dealing. The equations are, in fact, *linear* in the partial differential coefficients, and we begin by indicating an extension for them of the method used for ordinary differential equations with constant coefficients.

6. The equation of heat conduction. We return to the equation (p. 231)

$$\frac{\partial^2 V}{\partial x^2}=\frac{1}{h^2}\frac{\partial V}{\partial t},$$

and consider (analogously to the case of ordinary differential equations) whether there are solutions of the form

$$V=ae^{px+qt}.$$

If so, then
$$\frac{\partial^2 V}{\partial x^2}=ap^2e^{px+qt},\quad \frac{\partial V}{\partial t}=aqe^{px+qt},$$

so that
$$p^2=\frac{1}{h^2}q,$$

or
$$q=p^2h^2.$$

Hence the function
$$V=ae^{px+p^2h^2t}$$

satisfies the equation for all values of p. Further, it is verified by direct substitution that any sum of such solutions, for various values of p, is also a solution. Thus there exist solutions in the form

$$V=\sum_p A_p\,e^{px+p^2h^2t},$$

summed for any set of values of p.

For example, p might take in turn the complex values $i, 2i, 3i, \ldots, ni, \ldots$. The corresponding solution would be

$$V = \sum_{1}^{\infty} B_n \, e^{inx - n^2h^2t},$$

or, in real form,
$$V = \sum_{1}^{\infty} B_n \, e^{-n^2h^2t} \begin{cases} \sin nx \\ \cos nx \end{cases},$$

assuming that $B_1, B_2, B_3, \ldots$ are real constants.

7. Laplace's equation in three dimensions. For the equation

$$\frac{\partial^2 U}{\partial x^2} + \frac{\partial^2 U}{\partial y^2} + \frac{\partial^2 U}{\partial z^2} = 0,$$

we consider whether there are solutions of the form

$$U = a e^{px+qy+rz}.$$

If so, then the constants p, q, r must satisfy the condition

$$p^2 + q^2 + r^2 = 0,$$

and a form of solution is

$$U = \sum_{p,q,r} a_{pqr} \, e^{px+qy+rz} \qquad (p^2+q^2+r^2=0).$$

An obvious solution of the equation $p^2+q^2+r^2=0$ is given by $p = i\cos\alpha$, $q = i\sin\alpha$, $r = -1$. Then

$$U = \sum_{\alpha} b_\alpha \, e^{i(x\cos\alpha + y\sin\alpha) - z},$$

and the real and imaginary parts of this function are separate solutions of Laplace's equation. In real form we have, for example, a solution

$$U = \sum_{\alpha} c_\alpha \, e^{-z} \cos(x\cos\alpha + y\sin\alpha).$$

More generally, one at least of p, q, r must be complex, and the ranges of values to be selected depend, in any particular problem, on the type of solution that it is desired to obtain.

The great generality, which these solutions by sums of exponentials lead us to expect, prompts the question whether the exponential $e^{px+qy+rz}$ of the preceding paragraph may be replaced by a more general function $f(px+qy+rz)$ of the variable $px+qy+rz$. We therefore try such an assumption as a method for solving the wave equation in two dimensions.

8. The wave equation in a plane. Consider the equation

$$\frac{\partial^2 U}{\partial x^2}+\frac{\partial^2 U}{\partial y^2}=\frac{1}{c^2}\frac{\partial^2 U}{\partial t^2}$$

and solutions, if any, of the form

$$U=f(ax+by+\lambda t),$$

where a, b, λ are constants. Writing

$$u\equiv ax+by+\lambda t,$$

we have

$$\frac{\partial U}{\partial x}=\frac{\partial}{\partial x}f(u)$$

$$=\frac{df}{du}\frac{\partial u}{\partial x}$$

$$=a\frac{df}{du},$$

and, similarly,

$$\frac{\partial^2 U}{\partial x^2}=a^2\frac{d^2 f}{du^2}.$$

In the same way, we have

$$\frac{\partial^2 U}{\partial y^2}=b^2\frac{d^2 f}{du^2},\quad \frac{\partial^2 U}{\partial t^2}=\lambda^2\frac{d^2 f}{du^2},$$

and so the given equation is satisfied if a, b, λ are chosen so that

$$a^2+b^2=\lambda^2/c^2,$$

or

$$\lambda=\pm c\sqrt{(a^2+b^2)}.$$

Hence the equation is satisfied by the function

$$U=f\{ax+by+c\sqrt{(a^2+b^2)}\,t\}+F\{ax+by-c\sqrt{(a^2+b^2)}\,t\},$$

where f, F are arbitrary functions, and where the constants a, b are arbitrary.

An alternative generalization is to seek solutions which are functions of some *assigned* function of the variables. This is illustrated in the following section.

9. To examine solutions of Laplace's equation in the form of functions of $x+\sqrt{(x^2+y^2)}$. The equation is

$$\frac{\partial^2 U}{\partial x^2}+\frac{\partial^2 U}{\partial y^2}=0,$$

and the solution proposed is

$$U = f(x+r),$$

where $$r = \sqrt{(x^2+y^2)}.$$

For such a solution we have

$$\frac{\partial U}{\partial x} = f'(x+r)\left(1+\frac{\partial r}{\partial x}\right)$$

$$= \left(1+\frac{x}{r}\right)f'(x+r),$$

so that $$\frac{\partial^2 U}{\partial x^2} = \left(\frac{1}{r}-\frac{x}{r^2}\frac{x}{r}\right)f'(x+r)+\left(1+\frac{x}{r}\right)^2 f''(x+r)$$

$$= \frac{y^2}{r^3}f'(x+r)+\left(1+\frac{x}{r}\right)^2 f''(x+r).$$

Also $$\frac{\partial U}{\partial y} = f'(x+r)\frac{\partial r}{\partial y}$$

$$= \frac{y}{r}f'(x+r),$$

so that $$\frac{\partial^2 U}{\partial y^2} = \left(\frac{1}{r}-\frac{y}{r^2}\frac{y}{r}\right)f'(x+r)+\frac{y^2}{r^2}f''(x+r)$$

$$= \frac{x^2}{r^3}f'(x+r)+\frac{y^2}{r^2}f''(x+r).$$

The solution therefore satisfies the given equation if

$$\frac{1}{r}f'(x+r)+\left\{\left(1+\frac{x}{r}\right)^2+\frac{y^2}{r^2}\right\}f''(x+r)=0,$$

or $$\frac{1}{r}f'(x+r)+\left(1+\frac{2x}{r}+\frac{x^2+y^2}{r^2}\right)f''(x+r)=0,$$

or $$f'(x+r)+2(x+r)f''(x+r)=0.$$

If we write $x+r\equiv u$, then the form of the function f is given by the differential equation

$$f'(u)+2uf''(u)=0.$$

Hence $$u^{\frac{1}{2}}f'(u)=\tfrac{1}{2}A,$$

say, where A is an arbitrary constant; so that

$$f'(u)=\tfrac{1}{2}Au^{-\frac{1}{2}}$$

and $$f(u)=Au^{\frac{1}{2}}+B.$$

Hence a solution of Laplace's equation is obtained in the form

$$U = A\{x + \sqrt{(x^2+y^2)}\}^{\frac{1}{2}} + B,$$

where A, B are arbitrary constants.

The equations $\dfrac{\partial^2 U}{\partial x^2} + \dfrac{\partial^2 U}{\partial y^2} = 0$, $\dfrac{\partial^2 U}{\partial x^2} - \dfrac{1}{a^2}\dfrac{\partial^2 U}{\partial t^2} = 0$ may be solved by a simple transformation which gives a very general form of the result. But observe first that these two equations are identical in form (with t written for y) if the constant a is given the particular value $i \equiv \sqrt{(-1)}$. We therefore begin with the second equation.

10. To solve the equation $\dfrac{\partial^2 U}{\partial x^2} - \dfrac{1}{a^2}\dfrac{\partial^2 U}{\partial t^2} = 0$. If we make the substitution

$$u = x + at, \quad v = x - at,$$

then
$$\frac{\partial U}{\partial x} = \frac{\partial U}{\partial u} + \frac{\partial U}{\partial v}, \quad \frac{\partial^2 U}{\partial x^2} = \frac{\partial^2 U}{\partial u^2} + 2\frac{\partial^2 U}{\partial u\,\partial v} + \frac{\partial^2 U}{\partial v^2},$$

$$\frac{\partial U}{\partial t} = a\frac{\partial U}{\partial u} - a\frac{\partial U}{\partial v}, \quad \frac{\partial^2 U}{\partial t^2} = a^2\left(\frac{\partial^2 U}{\partial u^2} - 2\frac{\partial^2 U}{\partial u\,\partial v} + \frac{\partial^2 U}{\partial v^2}\right),$$

so that the equation
$$\frac{\partial^2 U}{\partial x^2} - \frac{1}{a^2}\frac{\partial^2 U}{\partial t^2} = 0$$

becomes
$$\frac{\partial^2 U}{\partial u\,\partial v} = 0.$$

Hence U assumes the form

$$U \equiv f(u) + g(v),$$

where f, g are arbitrary functions of their arguments. The solution of the given equation is therefore expressed by means of two arbitrary functions in the form

$$U = f(x+at) + g(x-at).$$

Note that, in this case, *all* solutions of the equation can be expressed in this form; in distinction from earlier cases, where special forms of possible solutions were guessed.

11. The solution of the equation $\dfrac{\partial^2 U}{\partial x^2} + \dfrac{\partial^2 U}{\partial y^2} = 0$ by means of conjugate functions. Putting $a = i$ ($i^2 = -1$) in the result of the preceding paragraph, we obtain a solution of Laplace's equation in two dimensions in the form

$$U = f(x+iy) + g(x-iy).$$

In particular, if we write

$$2f(x+iy) \equiv u(x,y) + iv(x,y)$$

and identify g with f, so that

$$2g(x-iy) \equiv u(x,y) - iv(x,y),$$

then we obtain solutions in the form

$$U = u(x,y).$$

Alternatively, if we identify g with $-f$, we obtain (after division by i) solutions

$$U = v(x,y).$$

Hence *solutions of the equation*

$$\frac{\partial^2 U}{\partial x^2} + \frac{\partial^2 U}{\partial y^2}$$

exist in the form of the real or imaginary part of the function $f(x+iy)$ of the complex variable $x+iy$.

12. The wave equation with spherical symmetry. The result of § 10 may be applied to find solutions of the equation

$$\frac{\partial^2 U}{\partial x^2} + \frac{\partial^2 U}{\partial y^2} + \frac{\partial^2 U}{\partial z^2} = \frac{1}{a^2}\frac{\partial^2 U}{\partial t^2}$$

having spherical symmetry, where U is a function of r, t only, with

$$r = \sqrt{(x^2+y^2+z^2)}.$$

In terms of spherical polars, the equation (p. 230) is

$$\frac{1}{r^2}\frac{\partial}{\partial r}\left(r^2\frac{\partial U}{\partial r}\right) + \frac{1}{r^2\sin\theta}\frac{\partial}{\partial\theta}\left(\sin\theta\frac{\partial U}{\partial\theta}\right) + \frac{1}{r^2\sin^2\theta}\frac{\partial^2 U}{\partial\phi^2} = \frac{1}{a^2}\frac{\partial^2 U}{\partial t^2},$$

and, since U is independent of θ and ϕ, this is

$$\frac{1}{r^2}\frac{\partial}{\partial r}\left(r^2\frac{\partial U}{\partial r}\right) = \frac{1}{a^2}\frac{\partial^2 U}{\partial t^2},$$

or

$$\frac{\partial^2 U}{\partial r^2} + \frac{2}{r}\frac{\partial U}{\partial r} = \frac{1}{a^2}\frac{\partial^2 U}{\partial t^2}.$$

Write
$$Ur \equiv V,$$
so that V is also a function of r, t only. Then
$$\frac{\partial U}{\partial r}r + U = \frac{\partial V}{\partial r},$$
$$\frac{\partial^2 U}{\partial r^2}r + 2\frac{\partial U}{\partial r} = \frac{\partial^2 V}{\partial r^2},$$
and so the equation is
$$\frac{\partial^2 V}{\partial r^2} = \frac{1}{a^2}r\frac{\partial^2 U}{\partial t^2} = \frac{1}{a^2}\frac{\partial^2 V}{\partial t^2}.$$
Hence (§ 10),
$$V = f(r+at) + g(r-at),$$
or
$$U = \frac{1}{r}\{f(r+at) + g(r-at)\},$$
where f, g are arbitrary functions of their arguments.

REVISION EXAMPLES XXIV

1. A string of length l is stretched between two points, one fixed and the other vibrating transversely. The motion of the string is determined by the equation
$$\frac{\partial^2 z}{\partial x^2} = \frac{1}{c^2}\frac{\partial^2 z}{\partial t^2},$$
where $z(x, t)$ is the transverse displacement at time t at a point at distance x from the fixed end, and c is constant. The motion of the end-points is given by
$$z(0, t) = 0, \quad z(l, t) = a \sin pt$$
for all t. Show that, in general, there is a solution in the form $z = f(x) \sin pt$, and determine $f(x)$.

In what circumstances does this solution fail?

2. For a transmission-line of uniform inductance L and capacitance C per unit length, the voltage V and current I satisfy the equations
$$\frac{\partial V}{\partial x} = -L\frac{\partial I}{\partial t}, \quad \frac{\partial I}{\partial x} = -C\frac{\partial V}{\partial t}.$$
Derive the partial differential equation satisfied by V or I, and show that there are solutions in the form of a function of x only,

multiplied by $e^{i\omega t}$, where ω is a given constant. Prove that the solutions are of the form

$$Ae^{i(\omega t-kx)}+Be^{i(\omega t+kx)},$$

where $k=\omega\sqrt{(LC)}$.

If $B=0$ for both V and I, show that the ratio of V to I is $\sqrt{(L/C)}$.

3. Write down a general solution of the wave equation

$$\frac{\partial^2 y}{\partial x^2}=\frac{1}{c^2}\frac{\partial^2 y}{\partial t^2}.$$

If the solution is subject to the boundary conditions that, for all t, $y=0$ when $x=0$ and when $x=l$, show that a solution of the form $y=f(x)\sin pt$ exists if, and only if, p has one of a series of discrete values.

4. Find the solutions of the wave equation

$$\frac{\partial^2 y}{\partial x^2}=\frac{1}{c^2}\frac{\partial^2 y}{\partial t^2}$$

for which $y=0$ at $x=0$ and at $x=l$ for all values of t.

Find the particular solution for which, in addition, at $t=0$,

$$y=0 \quad \text{and} \quad \frac{\partial y}{\partial t}=A\sin\left(\frac{3\pi x}{l}\right).$$

5. Expand $x(\pi-x)$ in a Fourier sine series in the range $(0,\pi)$.

Find the form and the coefficients of an infinite series of trigonometrical terms which represents, in the range $0\leqslant x\leqslant\pi$, a solution of the equation

$$\frac{\partial^2 y}{\partial x^2}=\frac{1}{c^2}\frac{\partial^2 y}{\partial t^2}$$

with the following boundary and initial conditions: (i) $y=0$ at $x=0$ and at $x=\pi$ for all t; (ii) $\partial y/\partial t=0$ and $y=x(\pi-x)$ when $t=0$ for all x in $(0,\pi)$.

6. The equation for the transverse vibrations of a stretched string is $\dfrac{\partial^2 y}{\partial x^2}=\dfrac{1}{c^2}\dfrac{\partial^2 y}{\partial t^2}$. The ends of the string being fixed at the points $(0,0)$, $(l,0)$, the string is released from rest in the form of an arc of the parabola $ay=x(l-x)$, where l/a is small. Show that its form at time t is given by the relation

$$ay=\Sigma\frac{8l^2}{\pi^3p^3}\sin\frac{p\pi x}{l}\cos\frac{p\pi ct}{l},$$

where p takes the odd values 1, 3, 5, ...

7. Obtain a solution of the equation

$$\frac{\partial z}{\partial y}=a\frac{\partial^2 z}{\partial x^2} \quad (a>0),$$

in the form $z=f(x)\,g(y)$.

Find the solution of this equation satisfying the conditions,

(i) z is finite as y tends to $+\infty$,

(ii) for all values of y, the value of $\partial z/\partial x$ is zero when $x=0$, and z is zero when $x=\frac{1}{2}\pi$,

(iii) for all values of x in the interval $(0, \frac{1}{2}\pi)$, the value of z is 1 when $y=0$.

8. A function V of x, y satisfies the equation

$$\frac{\partial^2 V}{\partial x^2}+\frac{\partial^2 V}{\partial y^2}=0$$

and vanishes for all values of y when $x=0$ and when $x=1$. Obtain an expression for V in the form $\Sigma f_n(y)\sin n\pi x$ satisfying the further conditions

(i) $V\to 0$ as $y\to+\infty$,

(ii) when $y=0$, the value of V is $\sin^3\pi x$ for $0\leqslant x\leqslant 1$.

9. Find all solutions of the differential equation

$$\frac{\partial^2 U}{\partial x^2}=\frac{\partial U}{\partial y}$$

which are of the form $U=f(x)\,g(y)$.

Solve the equation subject to the conditions that

$$U=0, \quad \frac{\partial U}{\partial x}=\cosh^2 y$$

when $x=0$, for all values of y.

10. Obtain a Fourier series, containing cosine terms only, for the function $f(\theta)$ defined by the relations

$$f(\theta)=1 \quad (0\leqslant\theta<\tfrac{1}{2}\pi),$$

$$f(\tfrac{1}{2}\pi)=0,$$

$$f(\theta)=-1 \quad (\tfrac{1}{2}\pi<\theta\leqslant\pi).$$

At time t, the excess pressure p at distance x from the closed end of a pipe of length a, open at the other end, may be assumed to satisfy the relations

$$\frac{\partial^2 p}{\partial x^2}=\frac{1}{c^2}\frac{\partial^2 p}{\partial t^2} \quad (0<x<a;\ \text{for all } t),$$

$$\frac{\partial p}{\partial x}=0 \quad \text{at} \quad x=0 \quad (\text{for all } t),$$

$$p=0 \quad \text{at} \quad x=a \quad (\text{for all } t>0).$$

If for $t \leqslant 0$, p has a constant value p_0 (the end of the tube at $x=a$ being closed, and then suddenly opened to the atmosphere at time $t=0$), show that the Fourier series found above enables us to express the subsequent variations of pressure in the tube by the formula

$$p=\sum_{k=1}^{\infty} A_k \cos\frac{k\pi x}{2a}\cos\frac{k\pi ct}{2a},$$

and find the coefficients A_k.

11. Find a Fourier sine series to represent the function

$$y=x \qquad (0 \leqslant x \leqslant a),$$

$$y=a \qquad (a \leqslant x \leqslant 2a),$$

$$y=3a-x \quad (2a \leqslant x \leqslant 3a).$$

What does the series represent in the interval $9a \leqslant x \leqslant 12a$?

Find a solution of the wave equation $\dfrac{\partial^2 y}{\partial t^2}=c^2\dfrac{\partial^2 y}{\partial x^2}$ which, when $t=0$, represents the foregoing function y, and also makes $\partial y/\partial t=0$ when $t=0$.

CHAPTER XXXI

SPHERICAL HARMONICS

The theory of spherical harmonics may be developed from several points of view, and is too extensive for more than a brief survey here. We seek to emphasize merely the properties to which they owe their special importance, and to sketch different treatments to which they may be subjected. The topics are selected chiefly for their use in applied mathematics, especially in the theory of electricity, with which we shall later assume that the reader has some acquaintance.

1. Laplace's equation for axial symmetry. When the potential U is symmetrical about an axis, taken to be the line $x=y=0$, Laplace's equation is (p. 230)

$$\frac{1}{r^2}\frac{\partial}{\partial r}\left(r^2\frac{\partial U}{\partial r}\right)+\frac{1}{r^2\sin\theta}\frac{\partial}{\partial\theta}\left(\sin\theta\frac{\partial U}{\partial\theta}\right)+\frac{1}{r^2\sin^2\theta}\frac{\partial^2 U}{\partial\phi^2}=0,$$

with (for the symmetry) U independent of ϕ; that is,

$$\frac{1}{r^2}\frac{\partial}{\partial r}\left(r^2\frac{\partial U}{\partial r}\right)+\frac{1}{r^2\sin\theta}\frac{\partial}{\partial\theta}\left(\sin\theta\frac{\partial U}{\partial\theta}\right)=0.$$

There exist (see p. 237) solutions of this equation in *separable* form

$$U=\sum_n\left(Ar^n+Br^{-(n+1)}\right)\Theta_n,$$

where Θ_n satisfies Legendre's differential equation

$$\frac{d}{d\mu}\left\{(1-\mu^2)\frac{d\Theta}{d\mu}\right\}+n(n+1)\,\Theta=0\quad(\mu=\cos\theta).$$

When n is a positive integer, Legendre's equation has a solution which is a polynomial of order n in μ, and it will be convenient for the calculations which follow to take the form (p. 182) in which the polynomial is expressed in *descending* powers of μ,

$$\mu^n-\frac{n(n-1)}{2(2n-1)}\mu^{n-2}+\frac{n(n-1)(n-2)(n-3)}{2.4.(2n-1)(2n-3)}\mu^{n-4}-\ldots,$$

or, after reduction,*

$$\frac{(n!)^2}{(2n)!}\sum_{\lambda=0}^{[\frac{1}{2}n]}\frac{(-1)^\lambda\,(2n-2\lambda)!}{(n-\lambda)!\,(n-2\lambda)!\,\lambda!}\mu^{n-2\lambda}.$$

2. The reciprocal of distance as a solution of Laplace's equation. The potential due to a single element (*point charge* in electrostatics or *particle* in gravitation) is inversely proportional to distance; thus the potential U at the point (x, y, z) distant s from an element at the point (x_1, y_1, z_1) is, apart from a coefficient of proportionality, given by the formula

$$U \equiv \frac{1}{s} = \frac{1}{\sqrt{\{(x-x_1)^2+(y-y_1)^2+(z-z_1)^2\}}}.$$

Since
$$s^2 = (x-x_1)^2+(y-y_1)^2+(z-z_1)^2,$$

it follows that
$$s\frac{\partial s}{\partial x} = x - x_1,$$

so that
$$\frac{\partial U}{\partial x} = -\frac{1}{s^2}\frac{\partial s}{\partial x} = -\frac{x-x_1}{s^3},$$

$$\frac{\partial^2 U}{\partial x^2} = -\frac{1}{s^3} + \frac{3(x-x_1)^2}{s^5}.$$

Thus
$$\begin{aligned}\nabla^2 U &= -\frac{3}{s^3}+\frac{3}{s^5}\{(x-x_1)^2+(y-y_1)^2+(z-z_1)^2\}\\ &= 0.\end{aligned}$$

This establishes the fundamentally important theorem:
The function s^{-1} satisfies Laplace's equation.

COROLLARY. The function

$$V \equiv \frac{\partial^{l+m+n}}{\partial x^l\,\partial y^m\,\partial z^n}\left(\frac{1}{s}\right),$$

where l, m, n are positive integers, also satisfies Laplace's equation, since

$$\begin{aligned}\nabla^2 V &\equiv \left(\frac{\partial^2}{\partial x^2}+\frac{\partial^2}{\partial y^2}+\frac{\partial^2}{\partial z^2}\right)\frac{\partial^{l+m+n}}{\partial x^l\,\partial y^m\,\partial z^n}\left(\frac{1}{s}\right)\\ &= \frac{\partial^{l+m+n}}{\partial x^l\,\partial y^m\,\partial z^n}\left(\frac{\partial^2}{\partial x^2}+\frac{\partial^2}{\partial y^2}+\frac{\partial^2}{\partial z^2}\right)\left(\frac{1}{s}\right)\\ &= 0.\end{aligned}$$

* We write $[\frac{1}{2}n]$ to denote $\frac{1}{2}n$ if n is even and $\frac{1}{2}(n-1)$ if n is odd.

Consider now a choice of coordinates in which the element is situated at the point $A(0,0,1)$, and suppose that the potential U is to be evaluated at the point $P(x,y,z)$ whose coordinates, when expressed in terms of spherical polars, are

$$x=r\sin\theta\cos\phi,\quad y=r\sin\theta\sin\phi,\quad z=r\cos\theta.$$

Then
$$s^2=1-2r\cos\theta+r^2,$$

so that
$$U=(1-2r\cos\theta+r^2)^{-\frac{1}{2}},$$

the positive determination of the square root being taken.

This is the function that we now investigate.

3. The expansion of $(1-2\mu r+r^2)^{-\frac{1}{2}}$. Writing μ for $\cos\theta$ in the formula at the end of the preceding section, we consider how the function
$$U\equiv(1-2\mu r+r^2)^{-\frac{1}{2}}$$

may be expanded in a series of ascending powers of r. As a preliminary step, we recall the binomial expansion, valid for $|x|<1$,

$$\begin{aligned}(1-x)^{-\frac{1}{2}}&=1+\tfrac{1}{2}x+\frac{\frac{1}{2}\cdot\frac{3}{2}}{2!}x^2+\frac{\frac{1}{2}\cdot\frac{3}{2}\cdot\frac{5}{2}}{3!}x^3+\dots\\&=\sum_{k=0}^{\infty}\frac{(2k)!}{2^{2k}(k!)^2}x^k.\end{aligned}$$

Since these coefficients occur several times in the calculations, we adopt the temporary notation
$$\alpha_k\equiv\frac{(2k)!}{2^{2k}(k!)^2},$$

so that the expansion is
$$(1-x)^{-\frac{1}{2}}=\sum_{k=0}^{\infty}\alpha_k x^k$$

for $|x|<1$.

Note, in particular, that *the coefficients α_k are all positive.*

In virtue of the identity

$$1-2\mu r+r^2\equiv(1-re^{i\theta})(1-re^{-i\theta}),$$

we have the relation

$$\begin{aligned}U&=(1-re^{i\theta})^{-\frac{1}{2}}(1-re^{-i\theta})^{-\frac{1}{2}}\\&=(1+\alpha_1re^{i\theta}+\alpha_2r^2e^{2i\theta}+\dots)(1+\alpha_1re^{-i\theta}+\alpha_2r^2e^{-2i\theta}+\dots),\end{aligned}$$

valid if $|re^{i\theta}|<1$ and $|re^{-i\theta}|<1$, that is, if

$$|r|<1.$$

Since each series is absolutely convergent for $|r|<1$, the product when arranged in a series of ascending powers of r also converges to the sum U (p. 132), so that, for $|r|<1$,

$$\begin{aligned} U = 1 &+ (\alpha_1 e^{i\theta} + \alpha_1 e^{-i\theta})\, r \\ &+ (\alpha_2 e^{2i\theta} + \alpha_1^2 + \alpha_2 e^{-2i\theta})\, r^2 \\ &+ (\alpha_3 e^{3i\theta} + \alpha_1\alpha_2 e^{i\theta} + \alpha_1\alpha_2 e^{-i\theta} + \alpha_3 e^{-3i\theta})\, r^3 \\ &\qquad\qquad + \ldots\ldots\ldots\ldots\ldots\ldots\ldots\ldots\ldots\ldots \\ = 1 &+ (2\alpha_1 \cos\theta)\, r + (2\alpha_2 \cos 2\theta + \alpha_1^2)\, r^2 \\ &\qquad\qquad + (2\alpha_3 \cos 3\theta + 2\alpha_1\alpha_2 \cos\theta)\, r^3 + \ldots. \end{aligned}$$

The actual values of the coefficients are not important for our present purpose (we shall find a more convenient expression later), but this *form* shows at once that, since $\alpha_1, \alpha_2, \ldots$ are all positive *the coefficients attain their greatest values when $\theta = 0$, that is, when $\mu = 1$.*

We now retrace our steps to establish an alternative version of the expansion. The expression $(1-2\mu r+r^2)^{-\frac{1}{2}}$ may be expanded as a series of ascending powers of the function $(2\mu r - r^2)$ in the form

$$(1-2\mu r+r^2)^{-\frac{1}{2}} = \sum_0^\infty \alpha_k (2\mu r - r^2)^k$$

for such values of $2\mu r - r^2$ as satisfy the inequality

$$|2\mu r - r^2| < 1.$$

This inequality does not necessarily extend for values of r up to the limit $|r|<1$ of the previous expansion; indeed, it is not satisfied for $\mu=-1$, $r=\frac{1}{2}$. But it certainly holds for values of r such that

$$|2r|+|r^2|<1,$$

and this, again, holds if (taking a crude but obvious inequality as adequate)

$$|r|<\tfrac{1}{3}.$$

Under this condition the series of *positive terms*

$$\sum_{k=0}^\infty \alpha_k \{|2r|+|r^2|\}^k$$

converges absolutely. Hence, using binomial expansions for the terms in brackets, the series

$$\sum_{k=0}^{\infty} \alpha_k \left\{ \sum_{p=0}^{k} {}_kc_p \,|\, 2r \,|^p \,|\, r^2 \,|^{k-p} \right\}$$

converges, and, because of the absolute convergence, continues to converge when re-arranged as a series in ascending powers of r. But this is precisely the condition that the series

$$\sum_{k=0}^{\infty} \alpha_k \left\{ \sum_{p=0}^{k} {}_kc_p (2\mu r)^p \, (-r^2)^{k-p} \right\},$$

when re-arranged as a series in ascending powers of r, should converge absolutely for $|r| < \frac{1}{3}$. This result establishes the validity of the expansion, for $|r| < \frac{1}{3}$,

$$(1-2\mu r+r^2)^{-\frac{1}{2}} = \sum_{k=0}^{\infty} \alpha_k \left\{ \sum_{p=0}^{k} {}_kc_p (2\mu r)^p \, (-r^2)^{k-p} \right\}$$

when written as a series in ascending powers of r.

An explicit form for the series may be found by observing that, for any given k, the term in r^n arises when

$$p+2(k-p)=n,$$

so that

$$p=2k-n.$$

Hence the total coefficient of r^n is

$$\Sigma \alpha_k \, {}_kc_{2k-n} (2\mu)^{2k-n} \, (-1)^{n-k},$$

summed for those values of k for which, n being given, the binomial coefficient ${}_kc_{2k-n}$ has a meaning; and, since

$${}_kc_{2k-n} \equiv \frac{k!}{(2k-n)!\,(n-k)!},$$

those values are given by the inequalities

$$\tfrac{1}{2}n \leqslant k \leqslant n.$$

Identification of this sum with the polynomial solution already found for Legendre's equation is more easily obtained if we arrange it in descending powers of μ, writing

$$2k-n=n-2\lambda,$$

or

$$k=n-\lambda.$$

Then the coefficient of r^n is

$$\Sigma \alpha_{n-\lambda} \cdot {}_{n-\lambda}c_{n-2\lambda} \cdot (2\mu)^{n-2\lambda} (-1)^\lambda$$

summed for $$0 \leqslant \lambda \leqslant \tfrac{1}{2}n.$$

Inserting values for $\alpha_{n-\lambda}$ and ${}_{n-\lambda}c_{n-2\lambda}$, we have the expression*

$$\sum_{\lambda=0}^{[\frac{1}{2}n]} \frac{(2n-2\lambda)!}{2^{2n-2\lambda}\{(n-\lambda)!\}^2} \frac{(n-\lambda)!}{(n-2\lambda)!\,\lambda!} (-1)^\lambda 2^{n-2\lambda} \mu^{n-2\lambda}$$
$$\equiv \sum_{\lambda=0}^{[\frac{1}{2}n]} \frac{(-1)^\lambda (2n-2\lambda)!}{2^n (n-\lambda)!\,(n-2\lambda)!\,\lambda!} \mu^{n-2\lambda},$$

agreeing, apart from a factor of proportionality (depending on n, but not on λ), with the polynomial solution of Legendre's equation given on p. 250.

We therefore summarize the present position as follows:

The expression $(1-2\mu r+r^2)^{-\frac{1}{2}}$ may be expanded in a series of ascending powers of r, *convergent for* $|r|<1$, in the form

$$(1-2\mu r+r^2)^{-\frac{1}{2}} = 1 + rP_1(\mu) + r^2P_2(\mu) + \ldots + r^nP_n(\mu) + \ldots;$$

and examination of the expression for the more restricted range $|r| < \frac{1}{3}$ has sufficed to establish that the typical coefficient $P_n(\mu)$ is a polynomial of degree n in μ, given by the formula

$$P_n(\mu) \equiv \sum_{\lambda=0}^{[\frac{1}{2}n]} \frac{(-1)^\lambda (2n-2\lambda)!}{2^n (n-\lambda)!\,(n-2\lambda)!\,\lambda!} \mu^{n-2\lambda}.$$

Further we have recognized this series as the polynomial solution of Legendre's equation; so that we know it to satisfy the relation

$$\frac{d}{d\mu}\left\{(1-\mu^2)\frac{dP_n}{d\mu}\right\} + n(n+1)P_n = 0.$$

It follows that, for arbitrary values of the constants A, B, the function

$$(Ar^n + Br^{-(n+1)})\,P_n(\mu)$$

is a solution of Laplace's equation with symmetry about the z-axis; more generally, a solution exists in the form of the infinite series

$$\sum_{n=0}^{\infty} (A_n r^n + B_n r^{-(n+1)})\,P_n(\mu).$$

* For $[\frac{1}{2}n]$ see p. 250.

DEFINITION. The function $P_n(\mu)$ is called *the* LEGENDRE POLYNOMIAL *of order* n. When the context makes the meaning clear, the argument μ is often omitted and the polynomial denoted by the simple symbol P_n.

It is easy to obtain explicit formulae for the first few polynomials:

$$P_1(\mu)=\mu,$$

$$P_2(\mu)=\tfrac{1}{2}(3\mu^2-1),$$

$$P_3(\mu)=\tfrac{1}{2}(5\mu^3-3\mu),$$

$$P_4(\mu)=\tfrac{1}{8}(35\mu^4-30\mu^2+3).$$

The values $P_n(-1)$, $P_n(0)$, $P_n(1)$ are of importance, and can be calculated directly from the expression $(1-2\mu r+r^2)^{-\frac{1}{2}}$, as follows:

When $\mu=-1$, the relation is

$$(1+r)^{-1}=1+rP_1(-1)+r^2P_2(-1)+r^3P_3(-1)+\ldots,$$

so that
$$P_n(-1)=(-1)^n.$$

When $\mu=+1$, the relation is

$$(1-r)^{-1}=1+rP_1(1)+r^2P_2(1)+r^3P_3(1)+\ldots,$$

so that
$$P_n(1)=1.$$

When $\mu=0$, the relation is

$$(1+r^2)^{-\frac{1}{2}}=1+rP_1(0)+r^2P_2(0)+r^3P_3(0)+\ldots,$$

so that
$$P_n(0)=0 \quad (n \text{ odd})$$

$$P_n(0)=\frac{n!}{(\frac{1}{2}n!)^2}\left(\frac{-1}{4}\right)^{\frac{1}{2}n} \qquad (n \text{ even}).$$

NOTE. We proved earlier (p. 252) that the greatest value of $P_n(\mu)$ occurs when $\mu=1$, so that

$$|P_n(\mu)|\leqslant 1.$$

This accords with the result that the series

$$1+rP_1(\mu)+r^2P_2(\mu)+\ldots$$

is convergent for all values of r such that $|r|<1$.

4. Rodrigue's formula for $P_n(\mu)$. The polynomial expression given for $P_n(\mu)$ in the preceding paragraph may be written more simply in the form (Rodrigue's formula)

$$P_n = \frac{1}{2^n n!}\left(\frac{d}{d\mu}\right)^n (\mu^2-1)^n.$$

To prove this directly, write

$$(\mu^2-1)^n = \sum_{\lambda=0}^{n} {}_nc_\lambda (-1)^\lambda (\mu^2)^{n-\lambda}$$

and note that, for $\lambda \leqslant \frac{1}{2}n$,

$$\left(\frac{d}{d\mu}\right)^n (\mu^{2n-2\lambda}) = \frac{(2n-2\lambda)}{(n-2\lambda)!}\,\mu^{n-2\lambda},$$

so that*

$$\frac{1}{2^n n!}\left(\frac{d}{d\mu}\right)^n (\mu^2-1)^n = \frac{1}{2^n n!}\sum_{\lambda=0}^{[\frac{1}{2}n]} \frac{(-1)^\lambda n!}{\lambda!\,(n-\lambda)!}\,\frac{(2n-2\lambda)!}{(n-2\lambda)!}\,\mu^{n-2\lambda}$$

$$= \sum_{\lambda=0}^{[\frac{1}{2}n]} \frac{(-1)^\lambda (2n-2\lambda)!}{2^n (n-\lambda)!\,\lambda!\,(n-2\lambda)!}\,\mu^{n-2\lambda}$$

$$= P_n(\mu).$$

5. Laplace's expressions for $P_n(\mu)$ as definite integrals. The polynomial $P_n(\mu)$ may also be expressed in integral form. We prove that *the value of $P_n(\mu)$ is given by the relations*

$$P_n(\mu) = \frac{1}{\pi}\int_0^\pi \{\mu + i\sqrt{(1-\mu^2)}\cos u\}^n\, du,$$

and

$$P_n(\mu) = \frac{\pm 1}{\pi}\int_0^\pi \{\mu + i\sqrt{(1-\mu^2)}\cos u\}^{-(n+1)}\, du.$$

(The forms are only superficially complex. When λ is odd,

$$\int_0^\pi \cos^\lambda u\, du = 0,$$

and so the terms involving i as a factor in the binomial expansion are all zero.)

We begin by proving the formula

$$\int_0^\pi \frac{du}{1+k^2\cos^2 u} = \frac{\pi}{\sqrt{(1+k^2)}}.$$

* When $\lambda > \frac{1}{2}n$, the differential coefficients are zero.

The left-hand side is

$$2\int_0^{\frac{1}{2}\pi}\frac{du}{1+k^2\cos^2 u}=2\int_0^{\frac{1}{2}\pi}\frac{\sec^2 u\,du}{(\tan^2 u+1)+k^2}$$

$$=2\int_0^{\infty}\frac{dt}{t^2+(1+k^2)}$$

$$=\frac{2}{\sqrt{(1+k^2)}}\left[\tan^{-1}\frac{t}{\sqrt{(1+k^2)}}\right]_0^{\infty}$$

$$=\frac{\pi}{\sqrt{(1+k^2)}},$$

the justification of the 'infinity' in the transformation being straightforward. Now

$$\int_0^{\pi}\frac{du}{1+k^2\cos^2 u}=\frac{1}{2}\int_0^{\pi}\frac{du}{1+ik\cos u}+\frac{1}{2}\int_0^{\pi}\frac{du}{1-ik\cos u},$$

and, with $u=\pi-v$, the second integral on the right is

$$\frac{1}{2}\int_{\pi}^{0}\frac{-dv}{1+ik\cos v}=\frac{1}{2}\int_0^{\pi}\frac{dv}{1+ik\cos v}.$$

The two integrals on the right are thus equal, and so

$$\frac{1}{\sqrt{(1+k^2)}}=\frac{1}{\pi}\int_0^{\pi}\frac{du}{1+ik\cos u}.$$

Note that each side is *positive* for all values of k.

In particular, if

$$k\equiv\frac{r\sin\theta}{1-r\cos\theta},$$

the formula is

$$\frac{\lambda(1-r\cos\theta)}{\sqrt{(1-2r\cos\theta+r^2)}}=\frac{1}{\pi}\int_0^{\pi}\frac{(1-r\cos\theta)\,du}{1-r\cos\theta+ir\sin\theta\cos u},$$

where λ is $+1$ or -1 according as $1-r\cos\theta$ is positive or negative. Dividing by $1-r\cos\theta$, which is not zero for general values of θ, we have

$$\frac{\lambda}{\sqrt{(1-2r\cos\theta+r^2)}}=\frac{1}{\pi}\int_0^{\pi}\frac{du}{1-r(\cos-i\sin\theta\cos u)}.$$

Now assume r positive and take $r<1$. Since

$$|r(\cos\theta-i\sin\theta\cos u)|<r,$$

the integrand on the right-hand side, when expanded in a series of ascending powers of r by the binomial theorem, is absolutely convergent and also, by the 'M'-test (p. 119), uniformly convergent; it is therefore integrable term by term. Moreover, with $r < 1$, the sign of $1-r\cos\theta$ is positive, so that $\lambda = +1$. Thus

$$\frac{1}{\sqrt{(1-2r\cos\theta+r^2)}}=\frac{1}{\pi}\sum_0^\infty\int_0^\pi r^n(\cos\theta-i\sin\theta\cos u)^n\,du,$$

so that, from the coefficients of r^n,

$$P_n(\mu)=\frac{1}{\pi}\int_0^\pi(\cos\theta-i\sin\theta\cos u)^n\,du,$$

where $\mu\equiv\cos\theta$. Replacing $-i$ by $+i$, which, by the parenthesis at the start of this paragraph, does not affect the result,

$$P_n(\mu)=\frac{1}{\pi}\int_0^\pi\{\mu+i\sqrt{(1-\mu^2)}\cos u\}^n\,du.$$

Again, the relation

$$\frac{\lambda}{\sqrt{(1-2r\cos\theta+r^2)}}=\frac{1}{\pi}\int_0^\pi\frac{du}{1-r(\cos\theta-i\sin\theta\cos u)}$$

may be written in the form

$$\frac{\lambda}{r\sqrt{(1-2r^{-1}\cos\theta+r^{-2})}}$$
$$=\frac{-1}{\pi}\int_0^\pi\frac{1}{r(\cos\theta-i\sin\theta\cos u)}\frac{du}{1-r^{-1}(\cos\theta-i\sin\theta\cos u)^{-1}},$$

and we proceed to consider this identity for values of r such that $r > 1$, so that $r^{-1} < 1$, and also such that, for given θ,

$$|r^{-1}(\cos\theta-i\sin\theta\cos u)^{-1}| < 1.$$

[This inequality requires $r^2\cos^2\theta+r^2\sin^2\theta\,\cos^2 u > 1$, which is true for all u if, and only if,

$$r|\cos\theta| > 1.]$$

The left-hand side is

$$\frac{\lambda}{r}\sum_0^\infty r^{-n}P_n(\mu),$$

or

$$\lambda\sum_0^\infty r^{-(n+1)}P_n(\mu);$$

and the right-hand side, by argument similar to that used for the case $r < 1$, is

$$-\frac{1}{\pi}\sum_{0}^{\infty}\int_{0}^{\pi}\frac{du}{r(\cos\theta - i\sin\theta\cos u)}\{r^{-1}(\cos\theta - i\sin\theta\cos u)^{-1}\}^{n},$$

or
$$-\frac{1}{\pi}\sum_{0}^{\infty} r^{-(n+1)}\int_{0}^{\pi}(\cos\theta - i\sin\theta\cos u)^{-(n+1)}\,du.$$

Hence, equating coefficients of $r^{-(n+1)}$, we obtain the formula

$$\lambda P_n(\mu) = \frac{-1}{\pi}\int_{0}^{\pi}\frac{du}{(\cos\theta - i\sin\theta\cos u)^{n+1}},$$

or (changing the irrelevant sign of i)

$$\lambda P_n(\mu) = \frac{-1}{\pi}\int_{0}^{\pi}\frac{du}{(\cos\theta + i\sin\theta\cos u)^{n+1}}.$$

$$= \frac{-1}{\pi}\int_{0}^{\pi}\frac{du}{\{\mu + i\sqrt{(1-\mu^2)}\cos u\}^{n+1}}.$$

In order to resolve the ambiguity of sign, consider the inequality

$$r\,|\cos\theta| > 1.$$

If $\cos\theta > 0$, then $1 - r\cos\theta < 0$, so that (p. 257)

$$\lambda = -1.$$

If $\cos\theta < 0$, then $1 + r\cos\theta > 0$, so that

$$\lambda = +1.$$

To summarize,

$$P_n(\mu) = \frac{+1}{\pi}\int_{0}^{\pi}\frac{du}{\{\mu + i\sqrt{(1-\mu^2)}\cos u\}^{n+1}} \quad (\mu > 0),$$

$$P_n(\mu) = \frac{-1}{\pi}\int_{0}^{\pi}\frac{du}{\{\mu + i\sqrt{(1-\mu^2)}\cos u\}^{n+1}} \quad (\mu < 0),$$

the integrals not being convergent for $\mu = 0$.

6. The recurrence relations. There are a number of useful formulae connecting successive Legendre polynomials and their differential coefficients. They may be derived from a number of starting points, of which we select Rodrigue's formula.

To prove the recurrence relations

$$P'_n - \mu P'_{n-1} = nP_{n-1},$$

$$\mu P'_n - P'_{n-1} = nP_n,$$

$$(n+1)P_{n+1} - (2n+1)\mu P_n + nP_{n-1} = 0.$$

Write
$$w_n \equiv \frac{1}{2^n n!}(\mu^2-1)^n,$$

so that
$$P_n = \frac{d^n w_n}{d\mu^n}.$$

By direct differentiation,

$$\frac{dw_n}{d\mu} = \frac{1}{2^{n-1}(n-1)!}(\mu^2-1)^{n-1}\mu$$

$$= \mu w_{n-1}.$$

Differentiate this relation n times by Leibniz's theorem; then

$$\frac{d^{n+1}w_n}{d\mu^{n+1}} = \mu\frac{d^n w_{n-1}}{d\mu^n} + n\,.\,1\,\frac{d^{n-1}w_{n-1}}{d\mu^{n-1}},$$

so that
$$\frac{dP_n}{d\mu} = \mu\frac{dP_{n-1}}{d\mu} + nP_{n-1},$$

or
$$P'_n - \mu P'_{n-1} = nP_{n-1}.$$

Again, the relation
$$\frac{dw_n}{d\mu} = \mu w_{n-1}$$

is
$$\mu\frac{dw_n}{d\mu} = \{(\mu^2-1)+1\}w_{n-1},$$

or
$$\mu\frac{dw_n}{d\mu} = 2nw_n + w_{n-1};$$

so that, differentiating n times by Leibniz's theorem,

$$\mu\frac{d^{n+1}w_n}{d\mu^{n+1}} + n\,.\,1\,\frac{d^n w_n}{d\mu^n} = 2n\frac{d^n w_n}{d\mu^n} + \frac{d^n w_{n-1}}{d\mu^n},$$

or
$$\mu\frac{dP_n}{d\mu} + nP_n = 2nP_n + \frac{dP_{n-1}}{d\mu},$$

or
$$\mu P'_n - P'_{n-1} = nP_n.$$

Finally, if we solve for P_n', P_{n-1}' the two equations

$$P_n' - \mu P_{n-1}' = nP_{n-1},$$

$$\mu P_n' - P_{n-1}' = nP_n,$$

we have

$$(1-\mu^2)\,P_n' = nP_{n-1} - n\mu P_n,$$

$$(1-\mu^2)\,P_{n-1}' = n\mu P_{n-1} - nP_n.$$

Replace $n-1$ by n in the latter equation, and then equate the two values of $(1-\mu^2)\,P_n'$; thus

$$nP_{n-1} - n\mu P_n = (n+1)\,\mu P_n - (n+1)\,P_{n+1},$$

or

$$(n+1)\,P_{n+1} - (2n+1)\,\mu P_n + nP_{n-1} = 0.$$

7. The integral formulae. The two formulae which follow, also called the *orthogonality relations*, are of great importance in applications.

(i) *To prove that, if $m \neq n$, then*

$$\int_{-1}^{1} P_m P_n \, d\mu = 0.$$

The polynomials P_n, P_m satisfy Legendre's equation in n, m respectively, so that

$$\frac{d}{d\mu}\{(1-\mu^2)\,P_n'\} + n(n+1)\,P_n = 0,$$

$$\frac{d}{d\mu}\{(1-\mu^2)\,P_m'\} + m(m+1)\,P_m = 0,$$

and so

$$P_m \frac{d}{d\mu}\{(1-\mu^2)\,P_n'\} - P_n \frac{d}{d\mu}\{(1-\mu^2)\,P_m'\} + (n-m)\,(n+m+1)\,P_m P_n = 0.$$

Hence, using integration by parts,

$$(m-n)\,(n+m+1)\int_{-1}^{1} P_m P_n \, d\mu$$

$$= \Big[P_m(1-\mu^2)\,P_n'\Big]_{-1}^{1} - \int_{-1}^{1} P_m'(1-\mu^2)\,P_n' \, d\mu - \Big[P_n(1-\mu^2)\,P_m'\Big]_{-1}^{1} + \int_{-1}^{1} P_n'(1-\mu^2)\,P_m' \, d\mu$$

$$= 0,$$

since $1-\mu^2$ vanishes both for $\mu=1$ and for $\mu=-1$. Also $m \neq n$ and (with $m, n \geqslant 0$) $n+m+1 \neq 0$, so that

$$(m-n)(n+m+1) \neq 0.$$

Hence
$$\int_{-1}^{1} P_m P_n \, d\mu = 0 \quad (m \neq n).$$

(ii) *To prove that*
$$\int_{-1}^{1} P_n^2 \, d\mu = \frac{2}{2n+1}.$$

Multiply the recurrence relation (p. 260)

$$nP_n - (2n-1)\mu P_{n-1} + (n-1) P_{n-2} = 0$$

by P_n and integrate, using the preceding result (i). Then

$$n\int_{-1}^{1} P_n^2 \, d\mu = (2n-1)\int_{-1}^{1} \mu P_{n-1} P_n \, d\mu.$$

Multiply similarly the relation

$$(n+1) P_{n+1} - (2n+1)\mu P_n + nP_{n-1} = 0$$

by P_{n-1} and integrate. Then

$$(2n+1)\int_{-1}^{1} \mu P_n P_{n-1} \, d\mu = n\int_{-1}^{1} P_{n-1}^2 \, d\mu.$$

Hence
$$\begin{aligned}(2n+1)\int_{-1}^{1} P_n^2 \, d\mu &= (2n-1)\int_{-1}^{1} P_{n-1}^2 \, d\mu \\ &= (2n-3)\int_{-1}^{1} P_{n-2}^2 \, d\mu \quad \text{(similarly)} \\ &= \dots\dots\dots\dots \\ &= \int_{-1}^{1} P_0^2 \, d\mu \quad \text{(similarly)} \\ &= 2 \quad (\text{since } P_0 = 1),\end{aligned}$$

and so
$$\int_{-1}^{1} P_n^2 \, d\mu = \frac{2}{2n+1}.$$

COROLLARY.
$$\int_{-1}^{1} \mu P_{n-1} P_n \, d\mu = \frac{2n}{(2n+1)(2n-1)}.$$

It is also easy to prove from the recurrence relation, or from the fact that the integrand is an odd function of μ, that

$$\int_{-1}^{1} \mu P_n^2 \, d\mu = 0.$$

8. Expansion in terms of Legendre polynomials. Comparison with Fourier series suggests the possibility that it may be possible to express a given function $f(x)$ as a series of Legendre polynomials in the form

$$f(x) \equiv a_0 P_0(x) + a_1 P_1(x) + a_2 P_2(x) + \dots .$$

We content ourselves with a proof of the formal theorem that, *if a function $f(x)$* CAN *be so expressed in a series which is uniformly convergent in the interval $-1 \leqslant x \leqslant 1$, then the coefficients a_k are given by the formula*

$$a_k = (k + \tfrac{1}{2}) \int_{-1}^{1} f(x) P_k(x)\, dx.$$

Since $P_k(x)$ is bounded in the interval and the series is uniformly convergent, so also is the series

$$f(x) P_k(x) \equiv a_0 P_0(x) P_k(x) + a_1 P_1(x) P_k(x) + \dots,$$

which may therefore be integrated term by term from $x = -1$ to $x = +1$. Hence, from the orthogonality relations (p. 261),

$$\int_{-1}^{1} f(x) P_k(x)\, dx = a_k \int_{-1}^{1} \{P_k(x)\}^2\, dx$$

$$= \frac{2a_k}{2k+1},$$

which is the required formula.

EXAMPLE

Prove that this formula is always valid when $f(x)$ is a given *polynomial.*

9. The application of Legendre polynomials; general principles. We conclude by giving three illustrations to show how Legendre polynomials are used in physical problems. But first we enunciate, without proof, certain theorems which form the background against which the methods employed must be tested. We adopt somewhat informal wording deliberately, to emphasize that we have not the equipment necessary for precision.

(i) THE THEOREM OF UNIQUENESS. If a potential function U is (by any method) found

(*a*) to satisfy Laplace's equation;

(*b*) to have assigned values over certain boundaries;

(*c*) to have 'assigned discontinuities' (possibly zero) in $\partial U/\partial n$ over certain boundaries, where $\partial U/\partial n$ denotes the rate of change of U in the direction normal to the boundary;

(*d*) to vanish 'sufficiently rapidly' (if required) at infinity, then that function is uniquely determined.

(ii) THE 'AXIS' THEOREM. If a potential function for a problem with axial symmetry (axis $x=y=0$) is determined for points on the axis, in the form

$$\sum_{n=0}^{\infty}(A_n z^n + B_n z^{-(n+1)}),$$

then the potential at the point (r, θ, ϕ) is

$$\sum_{n=0}^{\infty}(A_n r^n + B_n r^{-(n+1)})\, P_n(\mu) \quad (\mu \equiv \cos\theta).$$

It is unlikely that anyone reading the present work will not have some knowledge of the elementary electrical principles involved, but we shall give a brief note when any doubt seems likely to arise.

ILLUSTRATION 1. *To find the potential outside an earthed conducting sphere of radius a in the presence of a point charge e at a point A distant f from its centre* ($f>a$).

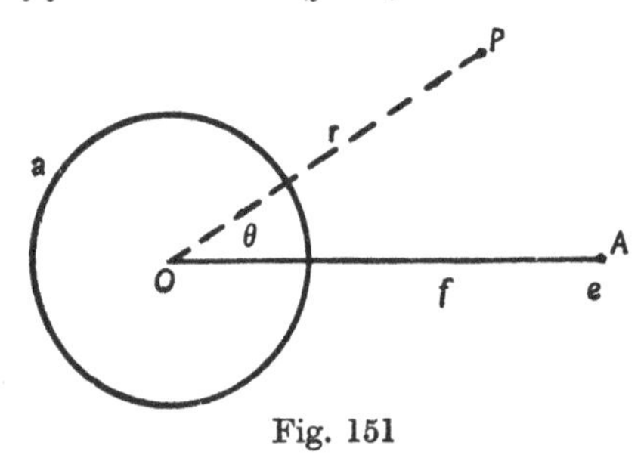

Fig. 151

The electrical principles are:

(i) the potential vanishes at all points on an *earthed* sphere;

(ii) the potential at a point P due to a point charge e at A is (in free space from which conductors, other charges, and so on, are absent) equal to e/AP;

(iii) the potentials at a point P due to two distinct effects may be *superposed*.

In terms of spherical polar coordinates, take the origin O at the centre of the sphere and the z-axis along OA, so that A is the point on it for which $z=f$. Let $P(r, \theta, \phi)$ be an arbitrary point outside the sphere.

We observe that, *in the absence of the sphere*, the potential at P is

$$\frac{e}{(f^2-2fr\mu+r^2)^{\frac{1}{2}}} \quad (\mu=\cos\theta).$$

Bearing in mind that we shall have to be considering the influence of the sphere $r=a$, where $a<f$, we form the expansion of this expression as a series of ascending powers of r, for values of r less than f. The expression is

$$\frac{e}{f}\left\{1-2\left(\frac{r}{f}\right)\mu+\left(\frac{r}{f}\right)^2\right\}^{-\frac{1}{2}},$$

so that the series is
$$\frac{e}{f}\sum_{n=0}^{\infty}\left(\frac{r}{f}\right)^n P_n(\mu).$$

The *total potential* U in space (outside the sphere) is the sum obtained by superposing the two components:

(i) $$\frac{e}{f}\sum_{n=0}^{\infty}\left(\frac{r}{f}\right)^n P_n(\mu) \quad (\text{for } r<f),$$

due to the point charge,

(ii) an expression, to be determined, due to the presence of the sphere.

In deciding what form to consider for the second of these components, we must bear in mind the facts that

(*a*) it satisfies Laplace's equation and has axial symmetry, so that it is the sum of terms of the type $(A_n r^n + B_n r^{-(n+1)})\,\Theta_n$ considered on p. 237;

(*b*) it tends to zero at infinity, far from the disturbing sphere;

(*c*) when added to the potential due to e at A, it gives zero potential over the sphere $r=a$.

Recalling the theorem of uniqueness (briefly, that if we can find A potential, then it is THE potential) we try the effect of choosing Θ_n of fact (*a*) to be the Legendre polynomial $P_n(\mu)$, thereby restricting n to the positive integers. Thus we try, in the first place, a 'disturbing' potential of the form

$$\sum_{n=0}^{\infty}(A_n r^n + B_n r^{-(n+1)})\,P_n(\mu).$$

Having done this, we note that, by (*b*), the coefficients A_n must all be taken to be zero, and so the total potential outside the sphere is

$$U=\frac{e}{f}\sum_{n=0}^{\infty}\left(\frac{r}{f}\right)^n P_n(\mu)+\sum_{n=0}^{\infty}B_n r^{-(n+1)}P_n(\mu).$$

Finally, we apply the condition (*c*), that $U=0$ for all values of μ when $r=a$. Thus

$$0 \equiv \frac{e}{f}\sum_{n=0}^{\infty}\left(\frac{a}{f}\right)^n P_n(\mu) + \sum_{n=0}^{\infty} B_n a^{-(n+1)} P_n(\mu).$$

This is satisfied (and we need not worry at this stage about convergence since only A solution is required; the convergence of any solution so proposed can be tested later) by choosing the coefficients B_n so that

$$\frac{ea^n}{f^{n+1}} + B_n a^{-(n+1)} = 0,$$

or

$$B_n = -\frac{ea^{2n+1}}{f^{n+1}}.$$

The disturbing potential is therefore found. It is (subject to convergence, which will be established almost immediately)

$$V \equiv -e\sum_{n=0}^{\infty}\frac{a^{2n+1}}{f^{n+1}}\frac{1}{r^{n+1}}P_n(\mu),$$

or

$$V = \left(-\frac{ea}{f}\right)\sum_{n=0}^{\infty}\left(\frac{a^2}{f}\right)^n \frac{1}{r^{n+1}}P_n(\mu).$$

Since $f > a$, this series certainly converges when $r \geqslant a$.

We have therefore proved that the potential outside the sphere assumes the form

$$U = \frac{e}{f}\sum_{n=0}^{\infty}\left(\frac{r}{f}\right)^n P_n(\mu) - \frac{ea}{f}\sum_{n=0}^{\infty}\left(\frac{a^2}{f}\right)^n \frac{1}{r^{n+1}}P_n(\mu)$$

for $r < f$; and the alternative expansion of the first summation for $r > f$ gives the form

$$U = e\sum_{n=0}^{\infty}\frac{f^n}{r^{n+1}}P_n(\mu) - \frac{ea}{f}\sum_{n=0}^{\infty}\left(\frac{a^2}{f}\right)^n \frac{1}{r^{n+1}}P_n(\mu)$$

for $r > f$.

Note that the *disturbing* potential is

$$-\frac{ea}{fr}\sum_{n=0}^{\infty}\left(\frac{a^2}{fr}\right)^n P_n(\mu)$$

$$= -\frac{ea}{fr}\left(1 - 2\frac{a^2}{fr}\mu + \frac{a^4}{f^2r^2}\right)^{-\frac{1}{2}}$$

$$= \frac{-ea/f}{\sqrt{\{r^2 - 2(a^2/f)\,r\cos\theta + (a^2/f)^2\}}},$$

which is, in fact, the potential due to a charge $-ea/f$ at the point B (inside the sphere) on the z-axis between O and A such that $OB = a^2/f$. The relation $OB.OA = a^2$ identifies B as the *inverse* of A with respect to the sphere.

Finally, we have obtained a potential which satisfies Laplace's equation, has the assigned value zero on the sphere, has a discontinuity of the same type as e/AP near the point A, and vanishes at infinity. It is therefore the unique potential which we sought.

ILLUSTRATION 2. *To find the potential inside and outside a sphere, of radius a, made of material of uniform dielectric constant K, when introduced into a field which, in its absence, was uniform of strength F.*

The electrical principles are:

(i) the potential U remains continuous as it crosses the surface of a dielectric;

(ii) the value of

$$-K\frac{\partial U}{\partial n}$$

(the negative sign being inserted merely to agree with the formula $-\partial U/\partial n$ for electric force in free space) remains continuous as it crosses a boundary on which there is no fixed charge;

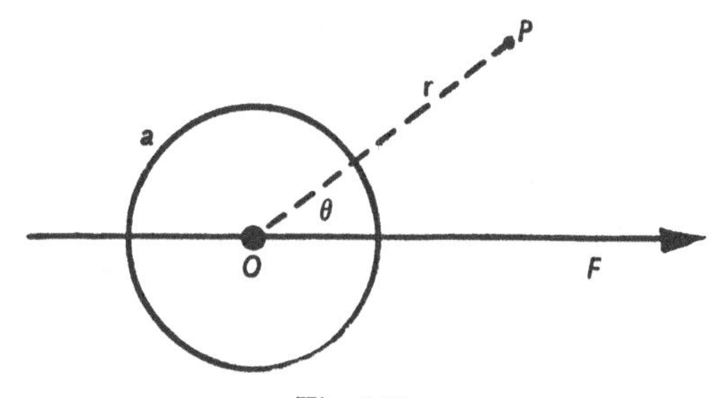

Fig. 152

(iii) the uniform strength F of the field in the absence of the dielectric implies that the potential U is then of the form

$$U_0 \equiv -Fz,$$

where the z-axis is taken in the direction of the field. (The value of U_0 may be increased by a constant if desired, but that has no essential effect on the problem.)

Thus, *in the absence of the sphere* the potential at P is

$$-Fr\cos\theta \equiv -FrP_1(\mu).$$

The *total potential* U in space will differ in algebraic form for points P inside or outside the sphere. In either case, it must be the sum of terms of the type

$$(A_n r^n + B_n r^{-(n+1)}) P_n(\mu)$$

for integral values of n. But for points *inside*, terms $B_n r^{-(n+1)}$ cannot be present, for the value $r=0$ would give an 'infinity'; and for points *outside*, terms $A_n r^n$ cannot be present (except for the given term $-Fr$) since the disturbance in potential due to the sphere must vanish at infinity. If, then, we denote the potentials inside and outside by the symbols U_i and U_o respectively, we have

$$U_i = \sum_0^\infty A_n r^n P_n(\mu),$$

$$U_o = -FrP_1(\mu) + \sum_0^\infty B_n r^{-(n+1)} P_n(\mu).$$

Since the potential is continuous at $r=a$, we have the identity

$$\sum_0^\infty A_n a^n P_n(\mu) \equiv -FaP_1(\mu) + \sum_0^\infty B_n a^{-(n+1)} P_n(\mu)$$

for all values of θ; and, since $-K\dfrac{\partial U}{\partial n}$ is continuous (that is, since $-K\dfrac{\partial U}{\partial r}$ is continuous, the normal being the radius) at $r=a$,

$$K\sum_0^\infty nA_n a^{n-1} P_n(\mu) \equiv -FP_1(\mu) - \sum_0^\infty (n+1) B_n a^{-(n+2)} P_n(\mu).$$

Equating coefficients of $P_n(\mu)$ in these identities, after multiplying the second throughout by a for convenience, we obtain the equations

$$\left.\begin{aligned} A_n a^n &= B_n a^{-(n+1)} \\ KnA_n a^n &= -(n+1) B_n a^{-(n+1)} \end{aligned}\right\} \quad (n \neq 1),$$

$$A_1 a = -Fa + B_1 a^{-2},$$

$$KA_1 a = -Fa - 2B_1 a^{-2}.$$

When $n \neq 1$, the relations can be satisfied only when $A_n = B_n = 0$. [With experience, this result may easily be foreseen and the formulae taken at once in the form $U_i = Ar\cos\theta$, $U_o = -Fr\cos\theta + Br^{-2}\cos\theta$.] When $n=1$, we have

$$A_1 = \frac{-3F}{K+2}, \quad B_1 = \frac{(K-1)a^3F}{K+2}.$$

Thus the potential assumes the form

$$U_i = -\frac{3Fr\cos\theta}{K+2},$$

$$U_o = -Fr\cos\theta + \frac{(K-1)a^3F\cos\theta}{(K+2)r^2}.$$

ILLUSTRATION 3. *To find the potential due to a ring of radius a charged with electricity to uniform density σ.* The electrical principles have already been enunciated.

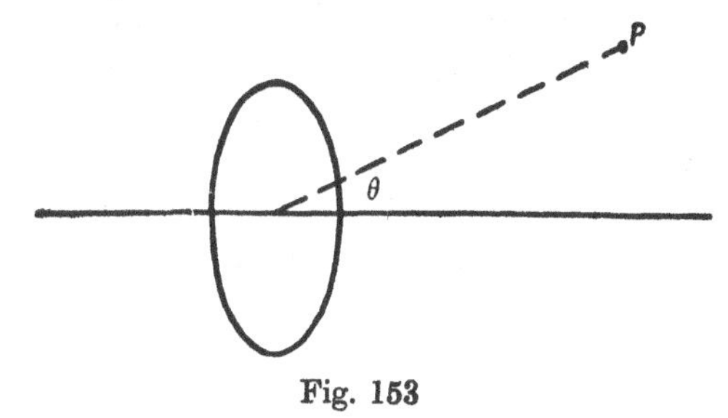

Fig. 153

We begin (in order to apply the 'axis' theorem enunciated on p. 264) by determining the potential at a point on the axis distant z from the centre of the ring. Since all points of the ring are distant $\sqrt{(a^2+z^2)}$ from that point, the potential there is

$$\frac{2\pi a\sigma}{\sqrt{(a^2+z^2)}}.$$

Thus
$$V_{z<a} \equiv 2\pi\sigma\left(1+\frac{z^2}{a^2}\right)^{-\frac{1}{2}}$$

$$= 2\pi\sigma\sum_{k=0}^{\infty}\frac{(2k)!}{2^{2k}(k!)^2}\left(-\frac{z^2}{a^2}\right)^k,$$

$$V_{z>a} \equiv 2\pi\left(\frac{a}{z}\right)\sigma\sum_{k=0}^{\infty}\frac{(2k)!}{2^{2k}(k!)^2}\left(-\frac{a^2}{z^2}\right)^k.$$

Hence, by the 'axis' theorem,

$$V_{r<a} = 2\pi\sigma\sum_{k=0}^{\infty}\frac{(2k)!}{2^{2k}(k!)^2}\left(-\frac{r^2}{a^2}\right)^k P_{2k}(\cos\theta),$$

$$V_{r>a} = 2\pi\sigma\sum_{k=0}^{\infty}\frac{(-1)^k(2k)!}{2^{2k}(k!)^2}\left(\frac{a}{r}\right)^{2k+1} P_{2k}(\cos\theta).$$

REVISION EXAMPLES XXV

1. Express Laplace's equation $\nabla^2 U = 0$ in terms of spherical polar coordinates. Show that, if U is of the form $R\cos\theta$, where R is a function of r only, then

$$R \equiv Ar + Br^{-2},$$

where A, B are constants.

2. Show that, in spherical polar coordinates, Laplace's equation $\nabla^2 U = 0$ has solutions $U = r\cos\theta$ and $U = r^{-2}\cos\theta$.

Fit these two solutions together to give a solution satisfying the following conditions: (i) $U \to 0$ as $r \to \infty$; (ii) U is finite at $r = 0$; (iii) U is continuous for all values of r; (iv) $\partial U/\partial r$ is continuous for all values of r except $r = a$; (v) $\partial U/\partial r$ increases discontinuously by an amount $k\cos\theta$ as r increases through the value a, where k is constant.

[Suggest, if you can, a physical interpretation of the solution.]

3. The density of charge σ at the point (a, θ, ϕ) on a spherical conductor of radius a placed in a certain electric field is given to be $\sigma \equiv k_1 P_1(\mu) + k_2 P_2(\mu)$, where $\mu \equiv \cos\theta$. The force acting on the conductor is known to be of magnitude $2\pi\sigma^2$ per unit area, acting outwardly along the radius. Prove that the resultant force is $32\pi^2 a^2 k_1 k_2/15$.

4. Evaluate (i) $\int_{-1}^{1} P_n(x)\,dx$, (ii) $\int_{-1}^{1} xP_n(x)\,dx$.

5. Prove that, if m, n are positive integers, both even or both odd, and $n \geqslant m$, then

$$\int_{-1}^{1} \frac{dP_m(x)}{dx}\frac{dP_n(x)}{dx}\,dx = m(m+1).$$

6. Evaluate

$$\int_{-1}^{1} x(1-x^2)\,P_n'(x)\,P_m'(x)\,dx.$$

7. If $P_n(x)$ satisfies Legendre's equation

$$(1-x^2)\,P_n'' - 2xP_n' + n(n+1)\,P_n = 0,$$

show that $V \equiv \dfrac{d^m P_n}{dx^m}$ satisfies the equation

$$(1-x^2)\,V'' - 2(m+1)\,xV' + \{n(n+1) - m(m+1)\}\,V = 0,$$

and that $$W \equiv (1-x^2)^{\frac{1}{2}m} \frac{d^m P_n}{dx^m}$$

satisfies $$(1-x^2) W'' - 2xW' + \left\{n(n+1) - \frac{m^2}{(1-x^2)}\right\} W = 0.$$

Hence show that $$\sin\theta \frac{dP_n(\cos\theta)}{d(\cos\theta)}$$

is a solution of

$$\frac{d^2 W}{d\theta^2} + \frac{d}{d\theta}(W \cot\theta) + n(n+1) W = 0.$$

8. Prove that

$$\int_{-1}^{1} x^n P_n(x)\, dx = 2^{n+1}(n!)^2/(2n+1)!.$$

9. Prove that

$$xP_n'(x) = nP_n(x) + (2n-3) P_{n-2}(x) + (2n-7) P_{n-4}(x) + \ldots.$$

10. Prove that

$$(x^2-1) P_n'(x) = nxP_n(x) - nP_{n-1}(x).$$

11. Evaluate $$\int_{-1}^{1} (x^2-1) P_{n+1}(x) P_n'(x)\, dx.$$

12. Prove that, if $f(x)$ is a polynomial of degree n,

$$f(x) = \frac{1}{2} \sum_{r=0}^{n} (2r+1) P_r(x) \int_{-1}^{1} f(t) P_r(t)\, dt.$$

Prove that, if n is a positive integer,

(i) $P_{n+1}'(x) - P_{n-1}'(x) = (2n+1) P_n(x)$,

(ii) $(n+1) P_{n+1}(x) + nP_{n-1}(x) = (2n+1) xP_n(x)$,

(iii) $(n+1) \int_0^1 P_n(x)\, dx = P_{n-1}(0)$.

Prove that, if $0 < x < 1$,

$$\sum_{n=1}^{\infty} \left(\frac{4n-1}{2n}\right) P_{2n-2}(0) P_{2n-1}(x) = 1.$$

13. Evaluate $$\int_{-1}^{1} x^{n+2} P_n(x)\, dx.$$

14. Calculate

$$\int_{-1}^{1} xP_n(x) P_{n+1}(x)\, dx, \quad \int_{-1}^{1} P_2(x) P_{n-1}(x) P_{n+1}(x)\, dx.$$

15. Assuming the formula $\int_{-1}^{1} P_n^2\,dx = 2/(2n+1)$, or otherwise, prove that

$$\frac{dP_n}{dx} = (2n-1)\,P_{n-1} + (2n-5)\,P_{n-3} + \dots,$$

the series terminating with $3P_1$ or P_0.

16. Prove that

$$\int_{-1}^{1} (1-x^2)\left(\frac{dP_n(x)}{dx}\right)^2 dx = \frac{2n(n+1)}{2n+1}.$$

17. Show that the integral

$$\int_{-1}^{1} x^m P_n(x)\,dx$$

is zero unless $m \geqslant n$ and $m-n$ is an even integer $2k$, and that its value is then

$$\frac{2(2k+n)\,(2k+n-1)\dots(2k+1)}{(2k+2n+1)\,(2k+2n-1)\dots(2k+1)}.$$

[The recurrence formula

$$\int_{0}^{1} x^{2k}(1-x^2)^q\,dx = \frac{2q}{2k+2q+1}\int_{0}^{1} x^{2k}(1-x^2)^{q-1}\,dx \quad (q \geqslant 1),$$

may be assumed.]

18. The functions $F_n(x)$ are defined by

(i) $F_n(x)$ is a polynomial of degree n,

(ii) $F_n(1) = 1$,

(iii) $\int_{-1}^{1} F_m(x)\,F_n(x)\,dx = 0 \quad (m \neq n)$.

Prove, by induction or otherwise, that $F_n(x) = P_n(x)$.

Hence, using only the properties (i), (ii), (iii), construct $P_0(x)$, $P_1(x)$, $P_2(x)$.

ANSWERS TO EXAMPLES

CHAPTER XIX

Examples I:

1. $y = x + \frac{1}{3}x^3 + A$.
2. $y = \tan(x + A)$.
3. $\sin y = A \sin x$.
4. $y = Ae^{x+\frac{1}{2}x^2} - 1$.
5. $3y = (1+x)^3 - 1$.
6. $\sec y = x - 2$.
7. $y^{-1} = (1+x)e^{-x} + 1 - 2e^{-1}$.
8. $(1+x)(1+y) = 27$.

Examples II:

1. $\log(x^2+y^2) = 2\tan^{-1}(y/x) + A$.
2. $\log\{(x+1)^2 + (y+1)^2\} = 2\tan^{-1}\{(y+1)/(x+1)\} + A$.
3. $\log y = \dfrac{x}{y} + A$.
4. $\log x = \dfrac{2}{\sqrt{3}}\tan^{-1}\left(\dfrac{2y+x}{\sqrt{3x}}\right) + A$.
5. $(x+2y^2)(x-2y^2)^3 = A$.
6. $\log(y^6 + 2xy^3 + 3x^2) = 2\sqrt{2}\tan^{-1}\left(\dfrac{y^3+x}{x\sqrt{2}}\right) + A$.
7. $x + \tan y = A(x - \tan y)^3$.
8. $\log(e^{2x} + e^{2y}) + 4\tan^{-1}(e^{y-x}) = A$.

Examples III:

1. $x^4y = A$.
2. $xy + y = A$.
3. $\cos x \sinh y = A$.
4. $x \sin y + y \cos y = Ae^{-x}$.
5. $xy(x^2 - y^2) = A$.
6. $2xy + 3x^2y - y^3 = A$.

REVISION EXAMPLES XV

1. $xy^2 - 2y = A$.
2. $2x^3 + 3x^2y + y^3 = A$.
3. $ax^2 + 2hxy + by^2 + 2gx + 2fy + A = 0$.
4. $(x+y^2)(x-y^2-2)^7 = A$.
5. $x^3 + y^3 + 3(xy - x + y) = A$.
6. $(x-y-1)^2 = A(3x-4y-6)$.
7. $x + y = Ae^{y-x}$.
8. $x + y = Ae^{-2xy/(x+y)^2}$.
9. $y^2 - x - 2 = Ae^{-\frac{1}{2}y^2}$.
10. $(y - \frac{1}{2}x^2 - A)(y - Ae^{-x}) = 0$.

11. $(y - Ae^{\cos x})(y - Ae^{-\sin x}) = 0$.

12. $y = \tanh x$.

13. $kv^2 = g(1 - e^{-2kx})$.

14. $y = x^2/(A - x)$.

15. $y = 1/(A - x - \log x)$.

16. $xy(x - y) = A$.

17. $(y - x - 1)^2 = A(x + y - 1)^3$.

18. $2 \sinh y = (e - e^{-1}) \sqrt{(x^2 + 1)}$.

19. $x = e^{-(x+y)/y}$.

20. $x - y \cos x = A$.

21. $\sqrt{2} \tan^{-1}\left\{\dfrac{(3y-2)\sqrt{2}}{3x+5}\right\} = \log\{(3x+5)^2 + 2(3y-2)^2\} + A$.

22. $y \sin x + \frac{1}{4} \sin 2x - \frac{1}{32} \sin 4x - \frac{3}{8}x = A$.

23. $(x - y)^2 = Axy^2$.

24. $y \cos x = 1 + xe^x - e^x$.

25. $7xy^2 = 9(x + 2y)$.

26. $\log \sqrt{(x^2 + xy + y^2)} + 3^{-\frac{1}{2}} \tan^{-1}\{(2y + x)/x\sqrt{3}\} = A$.

27. $x^3 - 3xy^2 = c^3$.

28. $(x + y)^2 = 2(x - y) + A$.

29. $y^3 - 3x^2y = c^3$.

30. (i) $x^3(x^2 + y^2)^5 (x^2 - 5y^2) = A$, (ii) $r = A(1 - \cos\theta)$.

CHAPTER XX

Examples I:

1. $6e^{4x}$.

2. 0.

3. $4x \sin 2x - 16 \cos 2x$.

4. $3 \cos^3 x - 2 \cos x$.

5. -1.

6. $4x^3$.

7. $(y'' + 2y' + y)\, e^{2x}$.

8. $y''' e^x$.

9. $x^2y'' + 6xy' + 6y$.

10. e^x.

11. $-\sin^3 x - 3 \cos^2 x \sin x$.

12. $4e^{5x} + 7e^{3x} + 3e^x$.

REVISION EXAMPLES XVI

1. $y = \cos^2 x\,(A + \sin x - \frac{1}{3}\sin^3 x)$.

2. $2(x + 1)\, y = x^4 - 6x^3 + Ax^2$.

3. $y = \left(\dfrac{x+1}{x-1}\right)^{\frac{1}{2}} \{\log(x + 1) + A\}$.

4. $y = -e^{-x\log x} + Ae^{x - x\log x}$.

5. $24x^3y\, e^x = 3(1 - 2x + 2x^2)\, e^{2x} - 4x^3 + A$.

6. $(2x+1)y = x(x+1)(x+3) + A(1+x)$.

7. $2y\log x = (\log x)^2 + A$.

8. $y\sin x = \frac{1}{2}x - \frac{1}{4}\sin 2x + A$.

9. $y = \frac{1}{2}(x+1)^3(x^2+2x+3)$.

10. $y = \frac{1}{5}\sin^4 x \tan x + C\sec x$.

11. $2(x+2)^2 y = (x+1)\{(x+1)^2 - 2\log(x+1) + A\}$.

12. $y(2-\cos x) = (2+\cos x)\{A - \log(2+\cos x)\}$.

13. $2y = (x+A)x^{\frac{1}{2}}\sin x$.

14. $y = x - 1 + A(x-1)e^{-\frac{1}{2}x^2}$.

15. $y = 1 - (x+A)/(\sec x + \tan x)$.

16. $(x-1)^3 y = (x+1)\{x^2 - 6x + 8\log(x+1) + A\}$.

17. $y = x(\log x + A)$.

18. $xy = \sin x - x\cos x - \pi$.

19. $y = 2\sin x + Ae^{-x}\cos x$.

20. $x^2 y = (x-1)\{\frac{1}{2}x^2 + 2x + 3\log(x-1) - (x-1)^{-1} + A\}$.

21. $y = (\sin^{-1}x)^2 + A\sin^{-1}x + B$.

22. $y = \frac{1}{4}x^2 - x + A\log x + B$.

23. $y = Bx^2 - A(1+x)$.

24. $y = (1+x^2)(A\cos x + B\sin x + x\sin x)$.

25. $y = Ax + Bx^2 - x\cos x$.

26. $y\sqrt{x} = A + B(1-x)^{a+1}$; $y\sqrt{x} = A + B\log(1-x)$.

27. $y = A(x^2+x) + Be^{2x} + \frac{1}{2}$.

28. $y = Ae^{-x} + B(x^2 - 2x + 2) + x^3 - 3x^2 + 6x - 6$.

29. $y = 1 + x^3\log x + Bx^3 + Ax^3\int x^{-4}e^{-\frac{1}{3}x^3}\,dx$.

30. $6xy = \sin x\left\{x^3 + B + A\int x^2\operatorname{cosec}^2 x\,dx\right\}$.

31. $y = Ax + B\cos x + \frac{1}{2}x\cos^2 x - \sin x\cos x$.

32. $axy = \pi\sin ax$.

33. $y^2 = 2x^2(\log x + A)$.

34. $B = A$, $n = 1$. $y = A(x + x^{-1}) + C\{(x + x^{-1})\tan^{-1}x - 1\} + x^2$

35. $2xy\sin x = x^3 + Ax + B$.

36. $y = x^4 + Ax^3 + Bx^{-3}$.

37. $\dfrac{dQ}{dx} + 3Q\sqrt{(2Q)} + 2PQ = 0$; $y = Ae^{\frac{1}{2}x^2} + Be^{x^2}$.

38. $y = \frac{1}{2}x + (Ax+B)e^{x^2}$.

39. $\{A+Bx^{1-2a}+x^2/(4a+2)\}e^{nx}$; $(A+B\log x+\frac{1}{4}x^2)e^{nx}$, when $a=\frac{1}{2}$, $(A+Bx^2+\frac{1}{2}x^2\log x)e^{nx}$, when $a=-\frac{1}{2}$.

40. $y=(A+Be^{-x})/x^2+x^2$.

41. $y(1+\sin x)=A\{\sin x+2\log(1-\sin x)\}+B$.

42. $y=A\cos x+B\tan x+\frac{1}{2}\cos x\tan^2 x$.

43. $(x+1)y=A+Bx^{n+1}+x^{n+2}/(n+2)$.

CHAPTER XXI

Examples I:

1. $y=Ae^x+Be^{2x}$.
2. $y=Ae^{-2x}+Be^{-\frac{1}{2}x}$.
3. $y=A\cos 3x+B\sin 3x$.
4. $y=(Ax+B)\cos x+(Cx+D)\sin x$.
5. $y=(A\cos 3x+B\sin 3x)e^{-4x}$.
6. $y=\left(A\cosh\frac{x\sqrt{5}}{2}+B\sinh\frac{x\sqrt{5}}{2}\right)e^{-\frac{1}{2}}$.
7. $y=(A\cos 4x+B\sin 4x)e^x$.
8. $y=(A+Bx+Cx^2)e^x$.
9. $y=(A+Bx+Cx^2)e^{-2x}$.
10. $y=(A\cos x+B\sin x)e^{-5x}$.

REVISION EXAMPLES XVII

1. $40y=e^{-2x}(3\cos 3x-e^{\pi}\sin 3x)+\sin 3x-3\cos 3x$.
2. $y=Ae^{-x}+Be^{-2x}+(\frac{1}{2}x^2-x)e^{-x}$.
3. $20y=15e^{-2x}-16e^{-3x}+e^{2x}$.
4. $y=Ae^{nx}+Be^{-nx}+xe^{nx}/(2n)$.
5. $y=(A\cos x+B\sin x+1)e^{-2x}$.
6. $y=e^{-\frac{1}{2}x}\left(\cos\frac{x\sqrt{3}}{2}+\sqrt{3}\sin\frac{x\sqrt{3}}{2}\right)-\cos x$.
7. $y=1-\frac{2}{3}\cos x-\frac{1}{3}\cos 2x$.
8. $y=(A\cos x+B\sin x)e^{-2x}+\frac{1}{65}(\sin 2x-8\cos 2x)$.
9. $y=(A+Bx)e^{-2x}+\frac{1}{4}+\frac{1}{8}\sin 2x$.
10. $144y=400e^{-2x}-261e^{-3x}+(12x+5)e^x$.
11. $y=(A\cos x+B\sin x)e^{3x}+2-\frac{1}{2}e^{2x}$.
12. $y=(2+3x)e^{-3x}+3x-2$.
13. $y=\sin x+\sin 2x+\cos 2x+2x+1$.

14. $y = Ae^{\frac{5}{2}x} + Be^{-\frac{1}{2}x} - xe^x$.

15. $y = (\sin x + 2\cos x)\,e^{-x} + (\sin x - 2\cos x)$.

16. $y = (A - \frac{1}{4}x)\sinh x + (B + \frac{1}{4}x^2)\cosh x$.

17. $y = (A + Bx - \sin x)\,e^{2x}$.

18. $50y = 53e^{-3x} + 155xe^{-3x} + 4\sin x - 3\cos x$.

19. $y = e^{-2x} + (x-1)\,e^{-x}$.

20. $x = (a + fn^{-2})\cos nt + \frac{1}{2}(t^2 - 2n^{-2})f$.

21. $x = ae^{-kt}\{\cos\sqrt{(n^2-k^2)}\,t + k(n^2-k^2)^{-\frac{1}{2}}\sin\sqrt{(n^2-k^2)}\,t\}$.

22. $y = \cosh x - 1$.

23. $y = e^{-x}(A + Bx + \frac{1}{4}x^2 + \frac{1}{8}\cos 2x)$.

24. $y = e^{2x}(A\cos x + B\sin x) + \frac{1}{125}(25x^2 + 40x + 22) + \frac{1}{65}(\cos 2x - 8\sin 2x)$.

25. $2y = (1+x)\,e^{-x} - \cos x$.

26. $y = \frac{1}{2}(\sin x - x\cos x)$.

27. $y = (2-x)\,e^x + e^{2x}$.

28. $y = e^x - 1$.

29. $y = Ae^x + B\cos x + C\sin x$.

30. $13x = (Ae^{-t} - 8)\cos 4t + (Be^{-t} + 1)\sin 4t$.

31. $y = (1-x)\,e^x$.

32. $27y = (A + Bx)\,e^{-3x} + 3x - 2$.

33. $10y = Ae^{-3x} + Be^{-x} + \sin x - 2\cos x$.

34. $40y = (Ae^{-2x} + 1)\sin 2x + (Be^{-2x} - 2)\cos 2x + 5$.

35. $4y = e^{-x}(A\cos x\sqrt{2} + B\sin x\sqrt{2}) + \cos x + \sin x$.

36. $24y = e^{\frac{3}{2}x}(Ax + B + x^3)$.

37. $y = Ae^{2x} + Be^{3x} + \sin x$.

38. $y = (A - 2x)\,e^{-2x} + Be^{-x} + \frac{1}{12}e^{2x}$.

39. $54y = A + Be^{-3x} - 2x + 3x^2 + 6x^3$.

40. $y = (A + Bx)\,e^{-2x} + \frac{1}{16}e^{2x}$.

41. $y = 2e^{3x} - 2e^k\sin 4x$, where $k = 3(\frac{1}{4}\pi - x)$.

42. $y = (A + \frac{1}{2}x)\sin x + B\cos x$.

43. $10y = Ae^{-2x} + Be^{-3x} - e^{-2x}(\cos 2x + 2\sin 2x)$.

44. $y = (A + Bx - \frac{1}{24}x^3)\sin x + (M + Nx - \frac{1}{8}x^2)\cos x + \frac{1}{18}\cos 2x + \frac{1}{2}$.

45. $y = (A + Bx + \frac{1}{2}x^2)\,e^{-x} + x - 2$.

46. $y = A + B\sin ax + C\cos ax - x^2a^{-2}\cos ax + 3xa^{-3}\sin ax$.

47. $y = Ae^{5x} + Be^x + Ce^{-4x} - \frac{1}{400}e^x(10x^2 + x) - \frac{1}{533}e^{-4x}(22\cos x - 7\sin x)$.

48. $y = A\sin 2x + B\cos 2x + x^2 \sin 2x.$

49. $y = Ae^{2x} + (B-3x)\,e^{x} + (C+x)\,e^{-x} + 6x + 3.$

50. (i) $x = 2e^{-t} + t - 2, \quad y = e^{-t} + 2t - 1$; (ii) $y(1) = e^{-1} + 1.$

51. $98x = 20e^{2t} - 20e^{-\frac{3}{2}t} + 28te^{2t},$
$98y = -49e^{-2t} - 11e^{2t} + 14te^{2t} + 60e^{-\frac{3}{2}t}.$

52. $x = \frac{1}{3}\cos t\sqrt{3} + \frac{2}{3}\sqrt{3}\sin t\sqrt{3} + \frac{2}{3} - \frac{1}{3}t + \frac{1}{6}t^2,$
$y = -\frac{7}{6}\cos t\sqrt{3} - \frac{1}{6}\sqrt{3}\sin t\sqrt{3} + 1 + \frac{1}{6}t^2 + \frac{1}{2}\sin t + \frac{1}{2}\cos t,$
$z = \frac{5}{6}\cos t\sqrt{3} - \frac{1}{2}\sqrt{3}\sin t\sqrt{3} + \frac{2}{3} + \frac{1}{3}t + \frac{1}{6}t^2 + \frac{1}{2}\sin t - \frac{1}{2}\cos t.$

53. $x = e^{-2t}(6t^3 - 3t^2 - 6t + 1), \quad y = e^{-2t}(-2t^3 + 3t^2 + 2t - 1).$

54. $x = Ae^{2t} + Be^{-t} + t, \quad y = -Ae^{2t} + 2Be^{-t} - 1.$

55. $13x = 4e^{2t} + 9e^{-\frac{1}{6}t}, \quad 13y = 13e^{t} - e^{2t} - 12e^{-\frac{1}{6}t}.$

56. $x = x_0\cos at + y_0\sin at + t\sin at,$
$y = y_0\cos at - x_0\sin at + t\cos at.$

57. $x = (1-t^2)\cos t, \quad y = 3\sin t + 14t\cos t - 3t^2\sin t.$

58. $2x = 3t(e^{t} - e^{-t}), \quad 2y = 3(3+t)\,e^{t} - (1-3t)\,e^{-t}.$

59. $y = e^{x} - \frac{1}{2}x, \quad z = -2e^{x} - \frac{1}{3}.$

60. $x = 4A\sin(t\sqrt{2} + \alpha) - B\sin(t\sqrt{5} + \beta) - \frac{1}{5},$
$y = B\sin(t\sqrt{5} + \beta) - A\sin(t\sqrt{2} + \alpha) + \frac{1}{5}.$

61. $x = 2\sinh t - 2t\cosh t, \quad y = -2\sinh t - 2t\cosh t - e^{2t}.$

62. $x = A\cosh(t+\alpha) - B\sin(t\sqrt{3} + \beta) - \frac{1}{16}\sinh t - \frac{1}{8}\sin t\sqrt{3}$
$+ \frac{1}{8}t\cosh t - \frac{1}{12}\sqrt{3}t\cos t\sqrt{3}.$
$y = A\cosh(t+\alpha) + 3B\sin(t\sqrt{3} + \beta) + \frac{3}{16}\sinh t - \frac{1}{8}\sin t\sqrt{3}$
$+ \frac{1}{8}t\cosh t + \frac{1}{4}\sqrt{3}t\cos t\sqrt{3}.$

63. $x = 4e^{t} + 11e^{7t} + 2t - 8, \quad y = 4e^{t} - e^{7t} + 3t - 15.$

64. $x = A\cos\frac{1}{2}(\sqrt{5}+1)(t+\alpha) + B\cos\frac{1}{2}(\sqrt{5}-1)(t+\beta) + \frac{1}{5}e^{t},$
$y = -A\sin\frac{1}{2}(\sqrt{5}+1)(t+\alpha) + B\sin\frac{1}{2}(\sqrt{5}-1)(t+\beta) + \frac{2}{5}e^{t}.$

65. $x = (A+Pt)\cos t - (B+Qt)\sin t - \frac{1}{2}t^2\sin t,$
$y = (B+Qt)\cos t + (A+Pt)\sin t + \frac{1}{2}t^2\cos t.$

66. $y = A\sin\log x + B\cos\log x + \frac{1}{2}x.$

67. $y = Ax^2 + Bx + \frac{1}{2}x^3.$ 68. $y = Ax^3 + Bx^2 + \frac{1}{2}x.$

69. $y = Ax^{-3} + Bx^{-4} + \frac{1}{2}x^{-3}\{(\log x)^2 - 2\log x\} + 21.$

70. $y = (Ax^{\sqrt{3}} + Bx^{-\sqrt{3}})\,x^2 + x^4.$ 71. $y = Ax^2 + Bx - x\cos x.$

72. $y = Ax + B\cos(\sqrt{3}\log x + \alpha) + x^2$.

73. $x = A + Bt + Ct^{-7} + \frac{1}{4}t\log t,\quad y = -\frac{7}{2}A - Bt + 7Ct^{-7} + \frac{3}{4}t - \frac{1}{4}t\log t$.

74. $x = A\sin(\alpha + \log t) + Bt + \frac{1}{2}t\log t + \frac{1}{5}t^2,$
$y = A\cos(\alpha + \log t) + Ct + \frac{2}{5}t^2 - \frac{1}{2}t\log t.$

75. $y = x^{-2}$.

76. $xy = Ae^{-1/x}$.

CHAPTER XXII

Examples I:

1. $y = \begin{cases} 0, \\ \frac{1}{2}x - \frac{3}{4} + \frac{1}{4}e^{-2(x-1)}, \\ \frac{1}{2} + \frac{1}{4}(1 - e^2)\,e^{-2(x-1)}. \end{cases}$

3. $a^2 y = 1 - (1 + at)\,e^{-at}$.

4. $y = e^x\{x\log x - x + 1\}$.

5. $y = a\cos x + b\sin x + x\sin x + \cos x\log(\cos x)$.

6. $N = \mu T^3/\{(\lambda + 2)(\lambda + 3)\}$.

7. $(R^2 + \omega^2 L^2)\,I/a = \begin{cases} L\omega\,e^{-Rt/L} + R\sin\omega t - L\omega\cos\omega t, \\ L\omega(1 + e^{R\pi/L\omega})\,e^{-Rt/L}. \end{cases}$

CHAPTER XXIII

Examples I:

1. Yes. 2. Yes. 3. No.
4. Yes. 5. Yes. 6. No.
7. Yes. 8. Yes. 9. Yes.
10. No. 11. Yes. 12. No.
13. Yes. 14. No. 15. Yes.

Examples II:

1. Yes. 2. Yes. 3. No.
4. Yes. 5. No. 6. Yes.
7. No. 8. No. 9. Yes.
10. No. 11. Yes. 12. Yes.
13. Yes. 14. Yes. 15. No.
16. Yes.

REVISION EXAMPLES XVIII

1. (i) $1-2^{-n}\to 1$; (ii) $\frac{1}{6}n(n+1)(2n+1)\to\infty$.

2. $\frac{3}{2}\{1-(-\frac{1}{3})^n\}\to\frac{3}{2}$.

3. (i) $\frac{1}{2}n(n+1)\to\infty$; (ii) $1-(n+1)^{-1}\to 1$.

4. $\frac{1}{2}$.

5. (i) $\dfrac{n}{3n+1}\to\dfrac{1}{3}$; (ii) $\frac{1}{3}n(n+1)(n+2)\to\infty$.

6. Not. ($u_n\to\pm\frac{1}{2}$).

7. $\dfrac{25}{48}-\dfrac{1}{4}\left(\dfrac{1}{n+1}+\dfrac{1}{n+2}+\dfrac{1}{n+3}+\dfrac{1}{n+4}\right)\to\dfrac{25}{48}$.

9. (i) $1-(n+1)^{-1}\to 1$; (ii) $\frac{1}{12}n(n+1)(n+2)(3n+17)$.

10. $4\sin A/(5-4\cos A)$.

11. $C=0$ $(\theta\neq k\pi)$; $S=\cot\theta$ when $\theta\neq k\pi$, $S=0$ when $\theta=k\pi$.

12. $\dfrac{3}{4}+\dfrac{1}{2(n+1)}-\dfrac{5}{2(n+2)}\to\dfrac{3}{4}$.

13. $\frac{1}{4}-S_n<\dfrac{1}{1,000,000}$ when $n>710$ approx.

15. The argument never holds.

19. All $\to 0$.

20. (i), (ii), (iv) Not sufficient by $1/n$; (iii), (v) sufficient, but not necessary by $1/n^2$.

22. Converges if $a\leqslant -1$.

26. (i) Diverges; (ii) converges if $k>2$.

28. 2.

29. Limit 2 if $u_1<3$; 3 if $u_1=3$.

32. Converges $\delta>0$.

36. $\log_e 8$.

37. $\log_e 3$.

38. Converges.

REVISION EXAMPLES XIX

3. No. (The limit is discontinuous at $x=1$.)

7. Not uniform—sum discontinuous at origin.

8. $p<2$.

9. Uniform.

10. Uniform for $p<\frac{1}{2}$.

12. (i) $p<1$, (ii) $p<2$.

14. $s(0)=1$, $s(x)=(1+x)^{-1}$ in $0<x<1$, $s(1)=1$.
Uniform in $0<x<1$.

REVISION EXAMPLES XX

1. ∞. 2. $\sqrt{2}$. 3. 1. 4. 1.
5. e. 6. 1. 7. $\frac{1}{2}$. 8. ∞.
9. 1. 10. 1. 11. 0. 12. $1/e$.
14. When both $|z| \geqslant \frac{1}{3}$ and $|z-3| \geqslant 1$.
15. (i) Necessary but not sufficient; (ii) not necessary, but sufficient.
16. Converges $|x|<1$. 17. Converges $|x|<1$.
18. (i) $0<x<2$, (ii) $|x|<1$. 19. $0<|x|<\sqrt{2}$.
20. All values. 21. Converges $|x| \leqslant 1/a$.
22. $\frac{1}{3}(2+3x-x^3)\log(1+x)-\frac{1}{3}(2-3x+x^3)\log(1-x)-\frac{4}{3}x+\frac{5}{9}x^3$.

REVISION EXAMPLES XXI

2. $\frac{1}{2}\pi$.
5. $\frac{1}{2}\pi(a+b)^{-1}$; $\frac{1}{2}\pi\log(1+b^{-1})$, $\pi\log 2$.
12. $\phi'(x)=\sqrt{\pi}$; integral $=\sqrt{\pi}$.
14. $-2(\sin^{-1}\alpha)^2$. 21. $\frac{1}{2}\pi\{1-\log\frac{1}{2}(e+1)\}$.
28. $a^{p+q-1}B(p,q)/\{b^q(b-a)^p\}$.

CHAPTER XXVII

Examples I:

1. General Bessel (p. 177) with $n=\frac{3}{2}$.
2. $$A\left\{1-\left(\frac{x}{2}\right)^2+\left(\frac{x^2}{2\,.\,4}\right)^2-\ldots\right\}+B\left\{\left[1-\left(\frac{x}{2}\right)^2+\ldots\right]\log x+\left(\frac{x}{2}\right)^2-\left(\frac{x^2}{2\,.\,4}\right)^2(1+\tfrac{1}{2})+\left(\frac{x^3}{2\,.\,4\,.\,6}\right)^2(1+\tfrac{1}{2}+\tfrac{1}{3})-\ldots\right\}.$$
3. $y=Au+B(\frac{1}{2}u\log x+v)$, where
$$u=x-\frac{x^3}{3^2-1}+\frac{x^5}{(3^2-1)(5^2-1)}-\ldots, \text{ and}$$
$$v=\frac{1}{x}+\tfrac{1}{4}x+\sum_1^\infty a_n x^{2n+1}, \text{ with}$$
$$a_n=\frac{(-)^n}{(3^2-1)(5^2-1)\ldots[(2n+1)^2-1]}\left\{\frac{1}{4}+\frac{3}{3^2-1}+\frac{5}{5^2-1}+\ldots\right\}.$$
4. $$Ax^{\frac{1}{2}}\left(1-\frac{x^2}{3!}+\frac{x^4}{5!}-\ldots\right)+Bx^{-\frac{1}{2}}\left(1-\frac{x^2}{2!}+\frac{x^4}{4!}-\ldots\right) \equiv Ax^{-\frac{1}{2}}\sin x+Bx^{-\frac{1}{2}}\cos x.$$

REVISION EXAMPLES XXII

1. $\dfrac{(-1)^{n-1}80x^n}{n!(3n-2)(3n-5)(3n-8)}$.

2. See p. 177.

3. The two series are

$$x^c\left\{1+\frac{c-3}{2c+1}(2x^2)+\frac{(c-3)(c-1)}{(2c+1)(2c+5)}(2x^2)^2 + \frac{(c-3)(c-1)(c+1)}{(2c+1)(2c+5)(2c+9)}(2x^2)^3+\dots\right\}$$

with $c=0, -1$, convergent for $|x|<1/\sqrt{2}$.

4. The two series are

$$x^c\left\{1+\frac{x}{c+1}+\frac{x^2}{(c+1)(c+2)}+\frac{x^3}{(c+1)(c+2)(c+3)}+\dots\right\}$$

with $c=0, \frac{1}{2}$.

5. $\Sigma a_n x^n$ is $1-x$;

$\Sigma b_n x^n$ is $1-\left(\dfrac{x}{1.3}+\dfrac{x^2}{3.5}+\dfrac{x^3}{5.7}+\dots\right)$.

7. The two series are

$$x^c\left\{1-\frac{c-3}{c+3}(2x^2)+\frac{(c-1)(c-3)}{(c+5)(c+3)}(2x^2)^2 - \frac{(c+1)(c-1)(c-3)}{(c+7)(c+5)(c+3)}(2x^2)^3+\dots\right\}$$

with $c=0, -1$, convergent for $|x|<\sqrt{2}$.

8. The series is $x(1+2x+3x^2+\dots)\equiv x(1-x)^{-2}$, convergent for $|x|<1$.
The second solution is $(x\log x+1)/(x-1)^2$.

9. (i) The two series are

$$x^c\left\{1+\frac{x}{(c+1)(c+k)}+\frac{x^2}{(c+1)(c+2)(c+k)(c+k+1)}+\dots\right\}$$

with $c=0$ or $1-k$.

(ii) $y=Au+B\left\{u\log x-2\sum_1^\infty \dfrac{x^\lambda}{(\lambda!)^2}\left(1+\frac{1}{2}+\dots+\frac{1}{\lambda}\right)\right\}$,

where $u=\sum_0^\infty \dfrac{x^\lambda}{(\lambda!)^2}$.

General solution when $k=\frac{1}{2}$ is $A\cosh(2\sqrt{x})+B\sinh(2\sqrt{x})$.

10. For arbitrary k, $y = Au + Bv$, where

$$u = 1 + \frac{x}{k} + \frac{x^2}{k(k+1)} + \dots,$$

$$v = x^{1-k}\left(1 + x + \frac{x^2}{2!} + \dots\right) \equiv e^x x^{1-k}.$$

11. $y = (A/x) + B\left\{-2x^{-2} + 1 + \tfrac{1}{4}x^2 + \frac{1.3}{4.6}x^4 + \frac{1.3.5}{4.6.8}x^6 + \dots\right\}.$

12. $y = A(x^2 + 6x + 12) + B\left\{x^5 + \frac{3}{1.6}x^6 + \frac{3.4}{1.2.6.7}x^7 + \frac{3.4.5}{1.2.3.6.7.8}x^8 + \dots\right\}$

13. $y = A\left(1 - \frac{x^6}{2!} + \frac{x^{12}}{4!} - \dots\right) + B\left(x^3 - \frac{x^9}{3!} + \frac{x^{15}}{5!} - \dots\right)$

$= A\cos(x^3) + B\sin(x^3).$

14. The two series are

$$x^c\left\{1 - \frac{(\lambda-4)+(c-2)^2}{(c+1)(c+2)}x^2 + \frac{[(\lambda-4)+(c-2)^2][(\lambda-4)+c^2]}{(c+1)(c+2)(c+3)(c+4)}x^4 - \dots\right\}$$

with $c = 0, 1$.

For $c = 0$, polynomial when $\lambda = 4 - (2\theta)^2$; for $c = 1$, polynomial when $\lambda = 4 - (2\theta+1)^2$, where, in each case, θ is an integer.

15. Particular solution is $u \equiv (x^2 - 4x + 2)e^{-x}\left[\equiv \frac{d^2}{dx^2}(x^2 e^{-x})\right]$. The general solution is $Au + Bv$, where

$$v \equiv u \log_e x + \sum_0^\infty (-)^n a_n x^n,$$

given by

$$a_n = \frac{3.4\dots(n+2)}{(n!)^2}\left\{\tfrac{1}{3} + \dots + \frac{1}{n+2} - 2\left(1 + \tfrac{1}{2} + \dots + \frac{1}{n}\right)\right\}.$$

16. $y = 1 + x + x^2 + x^3 + \dots.$

[The complete solution is $A(1-x)^{-1} + B(1-x)^{-1}\log x$.]

17. $y = A(1 + x + \tfrac{3}{4}x^2 + \tfrac{1}{2}x^3) + B(1 - \tfrac{1}{2}x)^{-2}.$

18. $$y = x + \frac{10(-7)}{3!}x^3 + \frac{10.12(-7)(-5)}{5!}x^5 + \frac{10.12.14(-7)(-5)(-3)}{7!}x^7 + \dots.$$

20. $y = Au + B(u \log x + 2xv)$, where

$$u = x + 2^2x^2 + \ldots + n^2x^n + \ldots,$$

$$v = \sum_1^\infty \left(\frac{n+1}{n}\right)^2 \left(\frac{1}{n+1} - \frac{1}{n}\right) x^n.$$

Radii of convergence 1.

REVISION EXAMPLES XXIII

1. $\sum_1^\infty (-)^{n-1} \dfrac{2n \sin \pi t}{\pi(n^2 - t^2)} \sin nx$; 0; $\sin t\,(x - 2\pi)$.

2. $$\frac{\sinh \pi}{\pi}\left[\frac{1}{2} + \sum_1^\infty \frac{(-)^n \cos nx}{1+n^2}\right] + \frac{1}{\pi}\sum_1^\infty \frac{n}{1+n^2}[1 + (-)^{n+1}\cosh \pi]\sin nx.$$

3. $$\frac{8}{\pi^2}\left[\sin x - \frac{\sin 3x}{3^2} + \frac{\sin 5x}{5^2} - \ldots\right],$$

$$f(x) = \frac{2(x - 2\pi)}{\pi} \qquad (2\pi \leqslant x \leqslant \tfrac{5}{2}\pi)$$

$$= 2 - \frac{2(x-2\pi)}{\pi} \qquad (\tfrac{5}{2}x \leqslant \pi \leqslant 3\pi).$$

5. $1 - \frac{1}{2}\cos x + 2\sum_2^\infty \dfrac{(-)^{n-1}}{n^2 - 1}\cos nx$. $(x - 2\pi)\sin x$.

6. $\sum_1^\infty \left[\dfrac{(-)^{n-1} 2\pi}{n} + \dfrac{4}{\pi n^3}\{(-1)^n - 1\}\right] \sin nx$. $-(2\pi - x)^2$.

7. $\frac{1}{3}\pi^2 + 4\sum_1^\infty \dfrac{(-)^n}{n^2}\cos nx$.

8. (a) $\dfrac{8}{\pi}\left\{\cos x + \dfrac{\cos 3x}{3^2} + \dfrac{\cos 5x}{5^2} + \ldots\right\}$; π; $\pi + 2x$.

   (b) $2\{\sin 2x + \frac{1}{2}\sin 4x + \frac{1}{3}\sin 6x + \ldots\}$; 0; $-\pi - 2x$.

9. $$\frac{2}{3} + \frac{\sqrt{3}}{\pi} \times \{\cos x - \tfrac{1}{2}\cos 2x + \tfrac{1}{4}\cos 4x - \tfrac{1}{5}\cos 5x + \tfrac{1}{7}\cos 7x - \tfrac{1}{8}\cos 8x + \ldots\};$$

   1, 0; sum $\pi/3\sqrt{3}$; $\frac{1}{2}$.

11. $\frac{1}{2}\pi + \dfrac{4}{\pi}\left(\cos x + \dfrac{\cos 3x}{3^2} + \dfrac{\cos 5x}{5^2} + \ldots\right)$.

12. $\frac{1}{6}a^2-(a/\pi)^2\left\{\cos\frac{2\pi x}{a}+\frac{1}{2^2}\cos\frac{4\pi x}{a}+\frac{1}{3^2}\cos\frac{6\pi x}{a}+\dots\right\}$; $\frac{1}{6}\pi^2$.

13. $\cos kx$; $\cos k(2\pi-x)$.

14. $\Sigma\,(2/n)\cos n\alpha \sin nx$.

15. $\Sigma\,(-)^{n-1}\left(\frac{2\pi^2}{n}-\frac{12}{n^3}\right)\sin nx$; $(x-2k\pi)^3$.

16. $\frac{2}{3}\pi^2-4\sum_1^\infty \frac{(-)^n}{n^2}\cos nx$.

17. $\frac{1}{3}\pi^2+4\sum_1^\infty \frac{(-)^n}{n^2}\cos nx$.

20. $\frac{1}{2}l-\frac{l}{\pi}\sum_1^\infty \frac{1}{n}\sin\frac{2n\pi t}{T}$.

22. $\sum_1^\infty\left\{\frac{\sin n\pi\epsilon}{n^2\pi^2\epsilon}+\frac{(-)^{n-1}}{n\pi}\right\}\sin n\pi x$.

In limit, z is $\frac{1}{2}$ for $0<x<1$ and $-\frac{1}{2}$ for $-1<x<0$. The series is

$$\frac{2}{\pi}\left\{\sin \pi x+\frac{\sin 3\pi x}{3}+\frac{\sin 5\pi x}{5}+\dots\right\}.$$

CHAPTER XXIX

Examples I:

4. $V=A\log\rho+B$, where ρ is distance from the line.

REVISION EXAMPLES XXIV

1. $f(x)=a\operatorname{cosec}\frac{pl}{c}\sin\frac{px}{c}\ \left(\frac{p}{c}\neq\text{integral multiple of }\frac{\pi}{l}\right)$.

2. $\frac{\partial^2 z}{\partial x^2}=LC\frac{\partial^2 z}{\partial t^2}$.

4. $y=\sum_k\left(P\cos\frac{k\pi ct}{l}+Q\sin\frac{k\pi ct}{l}\right)\sin\frac{k\pi x}{l}$; $y=\frac{Al}{3\pi c}\sin\frac{3\pi ct}{l}\sin\frac{3\pi x}{l}$.

5. $\frac{8}{\pi}\left\{\sin x+\frac{\sin 3x}{3^3}+\frac{\sin 5x}{5^3}+\dots\right\}$.

$$\frac{8}{\pi}\left\{\sin x\cos ct+\frac{\sin 3x\cos 3ct}{3^3}+\frac{\sin 5x\cos 5ct}{5^3}+\dots\right\}.$$

7. $\dfrac{4}{\pi}\{e^{-ay}\cos x - \frac{1}{3}e^{-9ay}\cos 3x + \frac{1}{5}e^{-25ay}\cos 5x - \ldots\}$.

8. $\frac{3}{4}e^{-\pi y}\sin \pi x - \frac{1}{4}e^{-3\pi y}\sin 3\pi x$.

9. $U = \frac{1}{2}x + \dfrac{\sqrt{2}}{8}\sinh(x\sqrt{2})\,e^{2y} + \dfrac{\sqrt{2}}{8}\sin(x\sqrt{2})\,e^{-2y}$.

10. $\dfrac{4}{\pi}\{\cos\theta - \frac{1}{3}\cos 3\theta + \frac{1}{5}\cos 5\theta - \ldots\}$.

$$A_k = \frac{4p_0}{k\pi}\sin\frac{k\pi}{2}.$$

11. $\dfrac{6a\sqrt{3}}{\pi^2}\left\{\sin\dfrac{\pi x}{3a} - \dfrac{1}{5^2}\sin\dfrac{5\pi x}{3a} + \dfrac{1}{7^2}\sin\dfrac{7\pi x}{3a} - \dfrac{1}{11^2}\sin\dfrac{11\pi x}{3a} + \ldots\right\}$,

where $b_n = \dfrac{12a}{n^2\pi^2}\sin\dfrac{n\pi}{2}\cos\dfrac{n\pi}{6}$.

$9a - x$ in $(9a, 10a)$; $-a$ in $(10a, 11a)$, $x - 12a$ in $(11a, 12a)$.

$y = \Sigma\, b_n \sin\dfrac{n\pi x}{3a}\cos\dfrac{n\pi ct}{3a}$ (b_n as before).

REVISION EXAMPLES XXV

2. $U_{r>a} = -\frac{1}{3}(ka^3/r^2)\cos\theta$, $U_{r<a} = -\frac{1}{3}kr\cos\theta$.
[Sphere with surface density $(-k/4\pi)\cos\theta$.]

4. (i) 0 when $n > 0$, 2 when $n = 0$;
(ii) 0 when $n > 1$, $\frac{2}{3}$ when $n = 1$, 0 when $n = 0$.

6. $I = 0$ unless $n = m - 1$ or $m = n - 1$. For the former,
$$I = \frac{2m(m^2-1)}{4m^2-1}.$$

11. $\dfrac{2n(n+1)}{(2n+1)(2n+3)}$.

13. $\dfrac{(n+2)!}{1.3.5.\ldots.(2n+3)}$.

14. $\dfrac{2(n+1)}{(2n+1)(2n+3)}$, $\dfrac{3n(n+1)}{(2n-1)(2n+1)(2n+3)}$.

INDEX